WYE ISLAND

Insiders, Outsiders, and Change
in a Chesapeake Community

Special Reprint Edition

WYE ISLAND

Insiders, Outsiders, and Change
in a Chesapeake Community

Special Reprint Edition

Boyd Gibbons

Resources for the Future
Washington, DC, USA

No part of this publication may be reproduced by any means, whether electronic or mechanical, without written permission. Requests to photocopy items for classroom or other educational use should be sent to the Copyright Clearance Center, Inc., Suite 910, 222 Rosewood Drive, Danvers, MA 01923, USA (fax +1 978 646 8600; www.copyright.com). All other permissions requests should be sent directly to the publisher at the address below.

An RFF Press book
Published by Resources for the Future
1616 P Street NW
Washington, DC 20036–1400
USA
www.rffpress.org

Library of Congress Cataloging-in-Publication Data

Gibbons, Boyd.
     Wye Island: insiders, outsiders, and change in a Chesapeake community / by Boyd Gibbons.—Special reprint ed.
          p. cm.
     Previously published: Harmondsworth, Eng.; New York: Penguin Books, 1979.
     ISBN 978-1-933115-42-9 (hardcover: alk. paper)—ISBN 978-1-933115-40-5 (pbk.: alk. paper)
     1. Land use—Maryland—Wye Island. 2. Rouse Company. 3. Real estate development—Maryland—Wye Island. I. Title.
     HD211.M3G5 2007
     333.78′30975232—dc22

The paper in this book meets the guidelines for permanence and durability of the Committee on Production Guidelines for Book Longevity of the Council on Library Resources.

The maps for this book were drawn by Frank and Clare Ford. This special reprint edition was typeset by Scribe, and the cover was designed by Henry Rosenbohm.

Cover illustration: Oil on linen painting, "Wye River Sailors," by Gavin Brooks, from the private collection of Mr. and Mrs. Michael Hankin.

The findings, interpretations, and conclusions offered in this publication are those of the author. They do not necessarily represent the views of Resources for the Future, its directors, or its officers.

ISBN 978-1-933115-42-9 (hardcover)   ISBN 978-1-933115-40-5 (pbk)

About Resources for the Future *and* RFF Press

Resources for the Future (RFF) improves environmental and natural re-source policymaking worldwide through independent social science research of the highest caliber. Founded in 1952, RFF pioneered the application of economics as a tool for developing more effective policy about the use and conservation of natural resources. Its scholars continue to employ social science methods to analyze critical issues concerning pollution control, energy policy, land and water use, hazardous waste, climate change, biodiversity, and the environmental challenges of developing countries.

RFF Press supports the mission of RFF by publishing book-length works that present a broad range of approaches to the study of natural resources and the environment. Its authors and editors include RFF staff, researchers from the larger academic and policy communities, and journalists. Audiences for publications by RFF Press include all of the participants in the policymaking process—scholars, the media, advocacy groups, NGOs, professionals in business and government, and the public.

CONTENTS

FOREWORD TO THE
SPECIAL REPRINT EDITION

When I started *Wye Island*, I thought the book might be something like John McPhee's *Encounters with the Archdruid*.

McPhee's classic appeared six years before *Wye Island*, in 1971. That was a year after the first Earth Day, when roughly 20 million Americans gathered to express concern about the degradation of the environment. The book brilliantly explored the dilemmas of development in an age of environmentalism. McPhee arranged for a leading environmentalist, David Brower, to meet three of his "natural enemies." In each case, the encounter took place on contested land. First, Brower hiked with mining engineer Charles Park in the Glacier Peak Wilderness, where the two debated whether the mountain's copper should be mined. Then Brower visited an almost uninhabited Georgia island with real-estate developer Charles Fraser, and the two debated Fraser's plan to build a resort there. Finally, Brower rafted the Colorado River through the Grand Canyon with the director of the U.S. Bureau of Reclamation, Floyd Dominy, who had overseen construction of most of the nation's biggest dams: The two debated plans, past and present, to build more dams on the river. Park, Fraser, and Dominy all were worthy antagonists. McPhee clearly appreciated many of their arguments, although he had great sympathy for Brower. The result was a book that still offers insight decades later.

In *Wye Island*, Boyd Gibbons similarly aimed to bring to life disparate views of the future of a place. Wye Island is on Maryland's Eastern Shore, and the book will have special meaning if you live or vacation on the Chesapeake. But the book transcends local interest. *Wye Island* is about why people fight over land use.

Like *Encounters with the Archdruid*, *Wye Island* is wonderfully well-written. That sort of skill always is remarkable, but especially so in this case: When Gibbons began the project, he had considerable experience in environmental policymaking, but he had never written a book. In the preface, Gibbons thanks McPhee for offering kind advice to "a novice." But Gibbons obviously was precocious. Though the phrase "land use" often seems lifelessly abstract, *Wye Island* enables readers to see the powerful passions, interests, and values that come into play in land-use controversies.

The controversy at the heart of the book dates from 1973, when the developer James Rouse began to plan a major complex on the island. Only a handful of people lived there. But many residents of nearby communities had a stake in the island's use. Rouse had built the celebrated "new town" of Columbia, Maryland, and he was one of the most respected developers in the nation. His reputation did not smooth his way. Like Fraser's proposed development in Georgia, Rouse's plans soon met resistance.

Gibbons was keen to put that resistance in social and historical context. He talked to members of the Watermen's Protective Association, who feared that Rouse's development would hurt the Chesapeake fisheries. He talked to people who lived nearby: Some were rich, some were retired, and almost all liked the island just the way it was. He talked to the three county commissioners, and he hung out in a general store owned by one of the three. He talked to the local developer who had invited Rouse to undertake the project. Because Gibbons was such a fine reporter and historian, readers come to appreciate Wye Island as a complex community, not simply a stage.

Gibbons did not shrink from exploring the class and racial tensions at work. Those tensions were most evident in discussion of the possible alternatives to Rouse's proposal. Though a few people imagined that the island might become a wildlife refuge, no one wanted a state park there. To some, the word "park" was something to spit out, with disgust. A park would attract black people from Baltimore, and that prospect was far more revolting than anything Rouse might do.

In the *Wye Island* cast of characters, however, no one played the Brower role. That surprised me. Individuals protested against the "rape" of the Bay and the "murder" of the Eastern Shore. But Gibbons cast doubt on their motives. Though some of the critics truly loved the island's wildlife, their deepest desire was not to protect the environment. Instead, they hoped to keep others away, to protect their privacy and their views of the water.

Who, then, spoke most powerfully for nature during the Wye Island controversy? Not every reader will answer that question the same way. My answer is the developer, James Rouse. Again, that surprised me. Rouse was drawn to the project for many reasons, but the most compelling—the reason he gave most often—was that he wanted to provide a model of ecologically sound development. In his initial letter to the Queen Anne's County commission, he promised that his project would embody "a new set of ideas that may lead the County, the state and the country to a higher understanding of how growth, water and the land can be more sensitively reconciled to one another." Later, when accused of raping and murdering the environment, he responded: "I can understand your fears and outrage. I can think of no image in America to which we can point as an adequate demonstration of what ought to be." Yet Rouse once again affirmed his belief "that there are new forms of development for ecologically sensitive land that can respect the land, the water, the fauna and the flora and accommodate rational, sensitive, imaginative development." Later still, Rouse made the point this way: "New kinds of large scale, well planned, good development are needed to provide images of the possible and to erect new standards to undergird stronger planning and zoning requirements and to make real to the people of a community that there are decent alternatives to sprawl and clutter and the ravaging of rivers."

Those words haunt me. Though Wye Island was not the best place to make a stand, Rouse was right about the need for positive models of development. Indeed, he understood one of the great needs of his time.

In the early 1970s, when Rouse offered his plan for Wye Island, the nation's policymakers were debating land-use legislation intended to encourage environmentally sensitive development. The legislation came after more than two decades of criticism of the environmental costs of suburban growth. Critics had decried the destruction of open space, the pollution of drinking water by failing septic tanks, and the folly of building in wetlands, floodplains, and steep hillsides, to name only three problems. Yet the effort to pass national land-use legislation eventually failed, in part because proponents could not illustrate their goal. As I argued in *The Bulldozer in the Countryside: Suburban Sprawl and the Rise of American Environmentalism*, the ideal of good land use was much harder to define than "clean air" or "clean water." Even where the air and water were polluted, people often could remember when the skies were blue or the streams were clear. But what was environmentally sound development? The reformers could not point to examples.

That still is a problem, unfortunately. In recent years, only one residential development has won national acclaim as a model of environmentally sensitive design: Prairie Crossing in Illinois. Why are models of good development so rare? *Wye Island* gave me some fresh ways of thinking about that question.

But you may draw other lessons from the book. Gibbons did not try to force conclusions on readers. He trusted that the story would be suggestive, and it is. The book's suggestiveness is part of its staying power. *Wye Island* will stick with you.

Adam Rome
Pennsylvania State University
March 2007

FOREWORD TO THE
ORIGINAL EDITION

When, in December 1973, an internal reorganization brought Boyd Gibbons into the division I was then heading, he seemed, above all else, a misfit. Not only was his subject matter unrelated to energy or materials—to the extent that there is anything unrelated to these two elements—but it seemed obvious that he was not writing a "scholarly" book. That is to say, he had resorted neither to statistical computations nor to a computer. He cited no literature and neither confirmed nor contradicted existing theories. There were no footnotes or references, no tables or graphs.

It took me some time to get on Boyd's wavelength; to understand, and more important, to appreciate, that his raw materials were people: their thoughts, emotions, prejudices, whims, and, eventually, their judgments coming together in collective decisions. At first I was disoriented and, in a way, disturbed by the seeming lack of logical order to the story he tells and the infeasibility of drawing conclusions that are generally applicable.

Out of a sense that a scholarly work must have "conclusions," I even drove him into an ill-fated attempt to distill findings out of his story so that it could stand up as a "project" rather than just a book. For that, I owe him a belated apology. The lesson is largely that there is no simple lesson and that the outcome cannot be fitted into a matrix, with all the variables labeled, numbered, and programmed.

Wye Island will take its place in the RFF publications list as evidence that "original research" is not found solely between the covers of a journal or scholarly tome. Boyd's love for his subject and his understanding of its inherent problems has enabled him to write an account that readers will find richly rewarding.

Hans H. Landsberg
Co-Director, Energy and
Materials Division
December 1976

PREFACE TO THE
ORIGINAL EDITION

For a number of years I had been thinking of exploring the connection between the ownership of land and the search for happiness. I was not entirely sure of what I was after, nor where such an inquiry would lead me. I knew only from where my interests had come—from impressions of the places I had lived. Los Angeles in the 1940s—on the way to the beach, past dairy farms and oil wells, and bomber plants lying under acres of camouflage netting. Driving cattle, belly-deep in snow, out of Montana's Bitterroot mountains. Vacant streets of a small town in Utah. Booming Arizona—orange groves and flower farms becoming, in the manner of Los Angeles, suburbs of Phoenix. Victorian turrets crumbling under the wrecking ball in Washington, DC. In these impressions of places and change lie the early roots of this book.

During the late 1960s and early 1970s, the environmental movement spread across the country and against the developers. Through new state laws, sewer moratoriums, and growth-control ordinances, citizens were

challenging the boosterism for development that, almost from the first colonial settlement, had characterized "progress" in America.

The stakes were high. Mayors and county commissioners feared that
their powers of land control would be usurped by governors and state legis-
latures, or, God forbid, by the federal government. Developers feared, at the
worst, bankruptcy and, at the least, costly governmental interference. Land
use was soon equated with the slowing down or stopping of growth, and
the passions now unleashed on all sides were seared and angry ones.

But the more involved I became in federal efforts to encourage state land
use laws, the less certain I was of what the goals ought to be, of where the
line should be drawn between the private and the public interest. I sensed
that most controversies over land development were steeped in personal
values. But what were these values? Were the voiced concerns the real concerns? What exactly were people trying to protect? The environment? Themselves? From what, and from whom? I was not sure. And I suspected that
many others shared my uncertainty.

The only way to understand these forces and attitudes, I felt, was to immerse myself in a particular dispute. The Eastern Shore of the Chesapeake
Bay was ideal, a scenic countryside attracting people from the cities across
the bay. For nine months I crisscrossed the upper Shore, gradually narrow-
ing my inquiry to Wye Island, where the highly respected Rouse Company
was proposing—against local opposition—a big residential development.
I dug into probate and land records, history books, old newspapers, scrap-
books, the files of the National Archives, and many other eclectic sources.
But mostly I just listened to and observed people wherever I found them.
On boats and tractors, in living rooms, on piers, and in general stores.

What follows is what I found there: how people behaved, what they
said, and how they felt. How typical is the Wye Island story and how universal its themes, I can only speculate. Certainly, we should be careful before applying too literally the lessons of a single place to others. But in
whatever way the particulars of this account differ from those elsewhere
where people seek to keep others out, the reader will, I suspect, find far
more similarities.

Although Resources for the Future has long been involved in studies of
land resources, this is not the sort of book RFF normally sponsors. Consequently, I am indebted to Joseph L. Fisher, then president of RFF (now

congressman from Virginia's Tenth District), for his willingness to try a different approach to policy research, and to Emery Castle, Hans Landsberg, Herb Morton, and Francis Christy here who urged its publication. Ruth Haas, Helen-Marie Streich, and Flora Stetson were most helpful with their continuing encouragement and support. I was fortunate in having an editor of Jo Hinkel's high caliber.

A great number of friends and associates were gracious with their time in reading and critiquing drafts of the manuscript. Among them I particularly want to thank Russell Train, Fred Bosselman, Steffen Plehn, Al Alm, Steve Jellinek, Kitty Gillman, John Fogarty, Bill Reilly, Terry Davies, Jack Noble, Bob Healy, Luther Carter, Checker Finn, Steve Hess, Neil Ward, Bob Eppstein, Steve Sloan, Schuyler Jackson, Henry Jarrett, Daniel Bromley, Bert Jacobson, Jack Schanz, and Cliff Russell.

Three others deserve special thanks for their contributions to this book: John Rosenberg, for helping me rework an overly burdened manuscript into clear and simple English; Ed Morgan, my good friend and coauthor of many unpublished U-boat scripts, who, besides commenting on early drafts, helped me keep my sense of humor; and John McPhee, for his kindness and advice to a novice about the craft of writing he masters so beautifully.

Finally, this book was only possible through the cooperation and assistance of hundreds of people on the Eastern Shore, in the Rouse Company, and elsewhere, some of whom appear in this book, many of whom do not. They are too numerous to mention by name, but it is to them I owe and hereby express my deepest gratitude.

Boyd Gibbons
Washington, DC
December 1976

WYE ISLAND

Insiders, Outsiders, and Change in a Chesapeake Community

Special Reprint Edition

It is a comfortable feeling to know that you stand on your own ground. Land is about the only thing that can't fly away. And then, you see, land gives so much more than the rent. It gives position and influence and political power, to say nothing of the game.

Anthony Trollope, The Last Chronicle of Barset

Every one of these hundreds of millions of human beings is in some form seeking happiness. . . . Not one is altogether noble nor altogether trustworthy nor altogether consistent; and not one is altogether vile. . . .

H. G. Wells, The Outline of History

CHAPTER 1

THE ROUSE VISION

From two hundred feet above the Baltimore ship channel, driving eastward over the Chesapeake Bay Bridge, James Rouse can see his native Eastern Shore stretching out ahead. It is March 14, 1974. Rouse has a plan for the Shore; it is in the trunk of his car. On an island tucked into the Shore, not far from the bay bridge, Rouse wants to build a waterfront village surrounded by large estates. Like the new town of Columbia, Maryland, that Rouse is building between Baltimore and Washington, DC, the Wye Island project is the blending of his dreams with his pragmatism. He believes that growth is inevitable on the Shore, but not the unplanned subdivisions that have crawled across the waterfront since completion of the bay bridge in 1952. Rouse believes that "good" development—not nondevelopment—is the answer to "bad" development. Jim Rouse is both a developer and a dreamer, and it is hard to distinguish where his dreams end and his business world begins. In taking on Wye Island, he is about to find out.

Maryland's Eastern Shore is part of the large, beet-shaped peninsula that separates the Chesapeake from Delaware Bay and the Atlantic. It encompasses, in addition to Maryland's Shore, the entire state of Delaware and a ragged sleeve of Virginia. When most native Marylanders speak of "the Shore," they are not usually referring to the beaches and high-rise condominiums of Ocean City, but of the land that faces Chesapeake Bay. This is flat country. The rolling dairy and grain farms at the head of the bay give way to more level ground. Once across the Choptank River (the rough division between the "upper" and "lower" Shore), the land slides like a spatula under the bay amid a vast fringe of salt marsh. The Chesapeake Bay is one of the world's largest estuaries, about two hundred miles from its head at the Susquehanna, the main source of its fresh water, to its mouth at Norfolk and the ocean. Its shoreline is as irregular as a tattered flag, shredded by thousands of necks, coves, and points of land, and its many sluggish creeks and rivers, like the Wye, giving the bay a circumference of amazing length. If stretched across the nation from coast to coast, its entire tidal shoreline would continue unraveling far out into both the Pacific and Atlantic oceans.

Jim Rouse is incurably sentimental about the upper Shore of his boyhood, its orchards and haystacks, workboats and ferries—and of small-town Easton where he grew up, which lies less than eight miles south of Wye Island. Rouse still clings to those memories, hoping to make them come true for others—at a profit. When Rouse left the Shore as an orphaned and almost penniless teenager during the Depression of the 1930s, he took the ferry to Baltimore, on the way to college, night law school, a career in mortgage banking, and finally, to his current position as chairman of the board and chief executive officer of the Rouse Company—overlord of Columbia and builder of huge shopping centers. But the ferries that took him away to fortune and prestige are gone, replaced by the bay bridge, which was built to get people to the ocean beaches beyond the Eastern Shore. Bridges for one purpose can become bridges for others. As the car leaves the bridge and rolls across Kent Island, Rouse can see the inverted metamorphosis of his dreams: billboards, real estate offices, and scattered subdivisions of houses.

Rouse drives over the stubby drawbridge at Kent Narrows, where oyster boats are growling out into Eastern Bay toward the tonging grounds. Soon

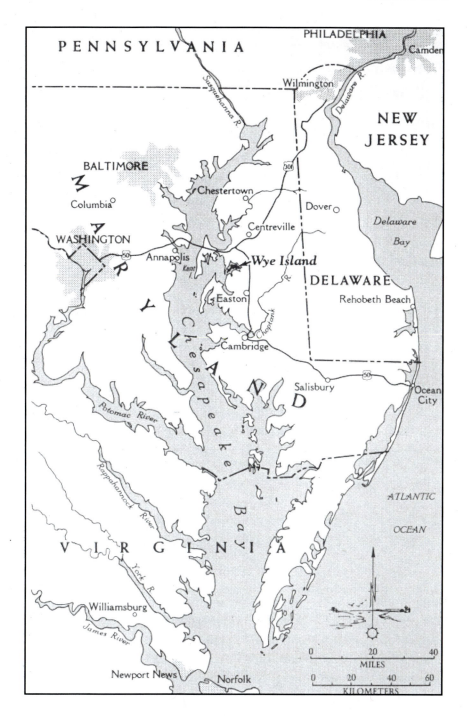

it will be the crabbing season, and the watermen will have to fight for water space with the summer yachts and weekend outboards. And everyone who lives near or works on the water will shudder when the first wave of "outsiders" sweeps down to the landings carrying their nets and twines baited with chicken necks, in pursuit of the Chesapeake blue crab. The *chickenneckers*, scourge of the Eastern Shore. Although the Eastern Shore is still rural, it is no longer remote.

Rouse is driving to Centreville, the seat of Queen Anne's County, where he hopes to convince the officials and the people of the county—both natives and newcomers—that he has their best interests at heart in his vision for Wye Island. Jim Rouse is in for a shock.

------ ∿ ------

The island that Rouse is planning to develop is—in the sense of geologic time—a youngster. Until about 500,000 years ago the Eastern Shore did not exist, except as the bottom of the sea. Then, when the earth cooled and glaciers absorbed much of its water, the seas fell, exposing large areas along the East Coast. During the warmer periods, the glaciers melted into rivers, which eroded the land and carried rocks and sediment seaward. One of these rivers was the Susquehanna, which meandered down a long valley. About 12,000 years ago enough of the ice cap had melted so that the rising seas flooded the valley, shortening the Susquehanna and creating the Chesapeake Bay. The waters climbed into smaller valleys along both shores of the bay, two of which are now the Wye and Wye East rivers. The land between these short parallel rivers (each roughly ten miles in length) would be a single neck, except that the waters rose high enough into the two channels to fill another channel—Wye Narrows—that crossed the end of the neck. Thus was Wye Island created.

The name Wye was probably affixed to the island and the rivers in the 1600s by Edward Lloyd, for the river of the same name in his native Wales. Lloyd settled directly across from the island, along what is now the Talbot County shore of the Wye East. He began to acquire land and to create descendants who would soon comprise one of the largest and most influential landholding families of Colonial Maryland. In the area around Wye Island the name of Wye has clung to no less than two rivers, a narrows, a neck of land, a landing, two towns (one long since vanished), the oldest and largest white oak in America, a gristmill, a church, four substantial estates and mansions, and, more recently, fourteen small businesses, including three gas stations, a gunning club, a feed company, an art gallery, and a firm of tree surgeons.

Wye Island, until recently, was relatively isolated. It has nurtured no village, nor is there a town of any size nearby. It lies on no main highway. It is simply at the end of a little-used county road, connected to Wye Neck across the narrows by a small bridge of old timbers. Like most of the Eastern Shore, Wye Island's 2,800 acres are large, flat fields of corn and soybeans, studded with woods of hickory, beech, and oak. Six miles long and almost forty miles around, Wye Island has changed little since the first English tobacco planters cleared its solid forest over three centuries ago. Wye Island reached its population peak (about a hundred people) between the latter part of the nineteenth century and the middle of the Depression, when it numbered thirteen farms of corn, wheat, vegetables, fruit orchards, turkeys, and some livestock. Most of these two dozen or so families were evicted in the twenties and thirties, when a couple of rich, eccentric recluses—the Stewarts—bought up two-thirds of the island, which was then fenced for cattle and patrolled by shotgun-carrying cowboys to keep people out. In the eyes of the natives who once lived there, Wye Island has gone downhill ever since.

As Jim Rouse drives into Centreville with his plans to save the Shore, Wye Island is all but uninhabited. It now contains only three farms of large cornfields and woods and hedgerows grown into high thickets. The small Bryan place, on the western tip, and the central two-thirds of the island owned by Frank and Bill Hardy, have only a few abandoned farmhouses. The eastern third of Wye Island—Wye Hall Farm—contains the island's entire population: Ruth and Harford Dorion, the farm's owners and infrequent visitors of their mansion, and the Bladeses and the Schuylers (the Dorlon's employees) who live in two frame houses nearby.

With his own portion of the island and an option on Wye Hall Farm (which includes all the island but the Bryan place), Frank Hardy approached Jim Rouse in the late 1960s, hoping to interest the Rouse Company in buying and developing Wye Island. But the company was busy building shopping centers and preoccupied in creating the centerpiece of Rouse's dreams: the new town of Columbia. Hardy had to wait. Then, in 1973, the Rouse Company was ready to replay the Columbia story elsewhere by developing smaller new towns, in the range of 10,000 to 25,000 people. A new communities division was formed, and after a lengthy search across the United States, two sites were chosen: Shelby Farms adjacent to Memphis, and Wye Island on Maryland's Eastern Shore. On May 1, 1973, people on the upper Eastern Shore awakened to the news that the Rouse Company had options to buy Wye Island. And Rouse was in trouble from the start.

The day that the Rouse option was announced, Rouse had appeared before the Queen Anne's County commissioners, Julius Grollman, Leonard Smith, and Jack Ashley, the governing body of the county. (Grollman and Smith each run small general stores, and Ashley is a farmer and in real estate.) The meeting was held in the commissioners' office in the courthouse at Centreville, the oldest courthouse still in use in Maryland. Rouse told the commissioners that he appreciated how they felt about trying to keep growth from overwhelming the county, and that the Rouse Company was willing to go to any length and spare no expense to design the finest possible, and most acceptable, development for the island; but, said Rouse, the final plan (then only a vague idea) would undoubtedly require a much higher density than its five-acre zoning allowed. Commissioner Leonard Smith told Rouse to do an impact study to show how the development would affect the county. Rouse said that the studies would be exhaustive, but the commissioners were unimpressed. They gave Rouse no encouragement for a later zoning amendment, or for anything else.

Later that summer Doug Godine (vice president for sales and marketing and Rouse's point man for the project) and Scott Ditch (an old Rouse hand and head of the company's public affairs division) made the first of their dozens of trips to the Shore to test local sentiment about the project. They met with the county commissioners. Godine described the studies already under way: the Rouse Company had assembled a team of Johns Hopkins University scientists to examine the hydrological and biological characteristics of the Wye River system around the island, contracted with two engineering firms to assess the road, sewer, and water supply problems, engaged a noted waterfowl expert to determine the feasibility of a wildlife sanctuary on the island, and had arranged with the land-planning and architectural firm of Wallace, McHarg, Roberts and Todd (WMRT) of Philadelphia for help in designing the plan.

Leonard Smith interrupted Godine: On second thought, forget those studies—we've just hired a planner and a public works director (the first in the county's history). Godine tried to sound positive. He said that the deeper the Rouse Company and its consultants probed into Wye Island's environment the more encouraged they were that they could create a unique, environmentally sound development, one that would cluster most people in a village rather than spreading them all over the island on five-acre homesites with hundreds of boat piers sticking into the rivers.

Julius Grollman, the president of the commissioners, looked at Doug Godine in the way that he might look through a window. "You've got a lot of convincing to do," he had said. "Wye Island is zoned R-1—for five-acre sites—and, as far as we're concerned, it's going to stay R-1."

"We would hope that you would at least listen," Godine replied.

"We will," said Grollman, "but we won't yield."

Unless there's enough money around to buy three new commissioners, Leonard Smith added, Wye Island will remain in five-acre zoning. It will not be changed. Not for Rouse, not for anybody. "It's quite a job," Smith had stated, "but we're not going to let that bay bridge change us at a rate we can't handle."

―――――― ᴄᴠɔ ――――――

Despite the immovable commissioners and the first blurred signals of a potential recession on the horizon, Jim Rouse enters Centreville on March 14, 1974, in an optimistic mood. He is convinced that his plan for Wye Island represents the only way the Shore can manage its growth without being inundated by it, that is, by concentrating development in a few places like Wye Island instead of sprawling it all over the waterfront. Since the announcement of the Rouse option in May 1973, the upper Eastern Shore has vibrated with rumors about precisely what the Rouse Company would build on Wye Island. Another Columbia? An amusement park? Rouse knows his proposal will be controversial, but he believes that in time he can convince the people of the county of its suitability.

Centreville looks much like it did at the turn of the century: Victorian homes on large lawns, a central courthouse, some shops, and quiet streets. You can walk from one end of town to the other in about fifteen minutes. A few buildings date from before the American Revolution. Rouse parks and walks into the county library to a meeting room where, shortly, he will

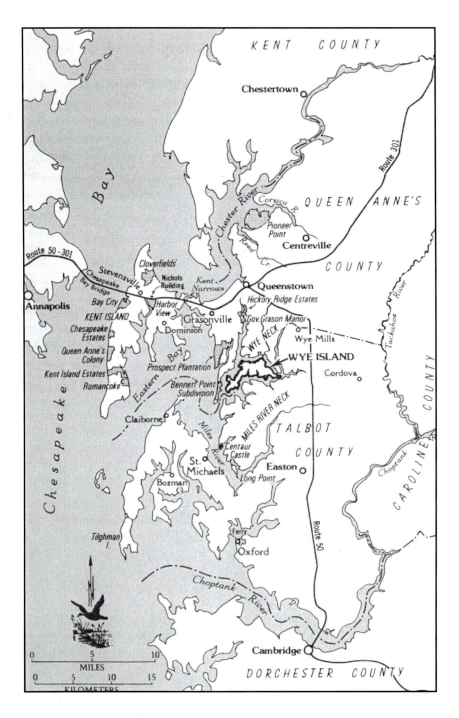

unveil his plan for Wye Island before the public and the county planning commission. He is expecting it to be a crowded, spirited meeting.

As people file in, Rouse stands in the center of the room, hands thrust in his trouser pockets. He blends easily into a crowd, for his physical appearance is unremarkable. There is no visual dazzle to James W. Rouse. He does not look a part of the intellectual circles and world of high finance and big business in which he moves; he does not sparkle with an expensive suit and banker's shoes. Rouse's outward manner is comfortable, quiet, and unhurried, that of an unpressed, balding, kindly professor, accustomed to penny loafers and rumpled sports jackets. His sister Dia—Lydia Pascault—fusses at his thinning hair with her pocket comb. She shakes her head and reminds her brother that he never pays enough attention to his appearance. Jim Rouse is sixty-one years old, a millionaire many times over, but he does not mind his sister's attention. He and Dia have always been very close, and her affectionate scolding is simply a reminder of what everyone knows about Rouse: appearances don't mean much to him. The Rouse Company has no company plane, and Rouse no limousine. Rouse's private office is almost public, on the ground floor of the Rouse Company building at Columbia. "Jim never puts on airs," says Dia. Rouse's idea of a good time is a pack trip into the Montana wilderness, or a weekend at his cabin on Long Point, which lies six miles south of Wye Island on the Miles River directly across from Dia's home.

Jim Rouse owns Long Point and for years camped out there. He built a log cabin on it, where he relaxes on whatever weekends his hectic business schedule allows. Dia's late husband O'Donnell was forever exasperated at Rouse, who would entertain family and important business associates (some still in business suits) by taking them over to Long Point in an open skiff. Often they would be caught in a storm on the water and tossed about and drenched. Rouse's nonchalance would exasperate O'Donnell. "Jim, you can't take people to Long Point like that," he would say, "you've got to have an adequate boat."

When Rouse finally relented, he contacted a local boat builder and asked the man to build him a solid workboat, like those used by the local watermen for oystering and crabbing. The boat is now moored at Dia's dock. Its name is painted across the stern: "ADEQUATE."

————— ✺ —————

The members of the planning commission gradually file into the crowded library and take their seats. The room quiets down and Robin Wood, the young planning administrator, begins by saying that the Rouse Company officials requested this opportunity to present, informally, their plan for Wye Island to the planning commission and the public, but that the company would not be requesting county approval at this time. Wood introduces Rouse, and Rouse begins to lay out the background of his firm's involvement in Wye Island. Rouse reads aloud a letter he had written to all the county residents the previous May, when the option was announced:

Our company has agreed to purchase 2,500 acres of Wye Island—one of Queen Anne's County's most important resources. You have a right to know our hopes and our intentions. We write to share them with you.

It is our firm belief that a quiet, beautiful county like Queen Anne's can grow in a manner that is consistent with its heritage in nature and in the basic serenity of its life. *We believe that the Wye River can be protected against pollution; that the oyster beds, the crabs and the fish can flourish; that the shoreline can be preserved to provide feeding grounds for ducks, geese and swan; that the farmland that marks the island's use can be significantly maintained* and that, at the same time, the island can become a place that supports a new waterfront village built to high standards of taste and quality unique in America.

We are acutely aware of the negative image that "development" evokes. We abhor the pattern of waterfront sprawl that has marked the pressing inroads of growth on the Eastern Shore.

We deeply, truly believe there is a way to manage growth that is sensitive to the land, the water and the people in the community. It will be our purpose to evolve a plan and proposal that gratifies the aspirations of the people of the County.

We believe that we can propose a future for Wye Island that will make it an even more important resource for the County than it has been in the past; that it can offer educational, cultural, aesthetic and economic benefits that will distinguish Queen Anne's County on the Shore and throughout the country.

To that end we will engage the best thinking available in the County and in the nation. We will be discussing our ideas and plans with you as they develop. We have no desire to undertake any proposal that does not have the enthusiastic support of the preponderance of the community. And we will not do so.

We write now to say that our purposes are where we believe your highest hopes would want them to be. We ask your patience while we plan and until we can propose a new set of ideas that may lead the County, the state and the country to a higher understanding of how growth, water and the land can be more sensitively reconciled to one another.

We solicit your ideas. Write or call us. We will soon have an office in Centreville, which we hope you will visit to talk with us about how you feel and what you think is best for the County.

Many thanks.

Sincerely,
James W. Rouse
Chairman of the Board

Rouse stands with his back to the window. To his right, resting on two long tables, is an immense scale model—eight by twelve feet in three abutted sections—of Wye Island and the surrounding rivers and countryside. Rouse suggests that the planning commission members gather about the model, to better understand the development scheme. Every chair is suddenly emptied in the rush toward the scale model. One man cranes far over the blue waters of the plastic Wye to scrutinize the miniature houses set out upon the island. Another picks at the tiny white sailboats near Drum Point. "Jesus, look at this detail," he says. Rouse waits for a moment. Then he reaches out across the island to the gate at the Bryan place and swirls his outstretched arm over Drum Point, up the island about two miles to its center, lifts it across to Shawn woods and out past Wye Hall Farm to the eastern end of the island. These two thousand acres, Rouse says, will be allowed for "estate residences" only, 184 parcels of fields, woods, and shoreline, ranging in size from five acres to over twenty. In the manner of a concert pianist striking a soft chord, Rouse drops his outstretched hand over Bigwood Cove, at the island's elbow.

"This will be Wye Village," he says.

Discord. The people squeezed together around the model flinch. Wye Village will have 706 dwelling units clustered at Bigwood and Grapevine coves. The total population of Wye Island will eventually reach about 2,750 people, Rouse says, most of them located in the village on only 12 percent of the island's surface. Wye Village will be on the scale of Oxford and Saint Michaels (two small waterfront towns in neighboring Talbot County), centering around a dock at Bigwood Cove and consisting of duplexes and apartments, a general store, some shops and offices for tradesmen, lawyers, and doctors, a small inn and conference center, and an eighteen-hole golf course. Rouse says that, although the sewage in the estate area will be handled by individual septic tanks, Wye Village—about 80 percent of the island's population—is to be served by an advanced wastewater treatment system, the Shore's first tertiary treatment plant. Its almost-pure effluent will be poured into ponds and sprayed onto the golf course and fields of Wye Island—not into the surrounding rivers.

Then, in what must seem like naiveté (or heresy) to those in the room who live in waterfront homes, Rouse announces that no homeowner on Wye Island will be allowed to build a private boat pier along the shoreline. All boat slips on Wye Island—limited to two hundred—will be concentrated at the village dock, and no more than twenty boats can be powered solely by engines. The onlookers crinkle their faces.

Rouse continues. No one who buys land on Wye Island will be allowed to resubdivide their property—ever. A woman suddenly laughs. To help arrest further erosion of the river banks, none of the owners of estate residences will be permitted to build within two hundred feet of the shoreline. An architectural panel must approve the design and construction of all homes. Almost half of the estate area, about eight hundred acres, will be maintained as farmland, and most of the five hundred wooded acres on Wye Island, particularly those anchoring the river banks, will remain untouched. Another four hundred acres of Wye Hall Farm will be protected as a wildlife trust, and new marshes will be created. In all, Rouse states, half of Wye Island is to remain as it is—undeveloped.

Rouse looks at the faces around the model, hoping to detect at least a shadow of interest. He sees only skepticism.

———————ᘐᕁᖇ———————

In the early evenings of summer four young boys would meet at the post office in Easton. From here they would set off, flying down the sidewalk toward home, trying to avoid being tagged by the one who was "it." The game was called last-tag. When Jim Rouse did the chasing, the other three would grit their teeth, because, despite his smaller size and a bout with infantile paralysis, Rouse could run quite fast. Of more concern to them, they knew he would persevere to the bitter end. Rouse once tagged a boy with sufficient authority to send him flying over a hedge, and on another occasion, he dogged Porter Mathews right up the steps into the Mathews's house and caught him at the door of his bedroom.

Until he was fifteen, Jim Rouse lived a relatively secure and pleasant life. He was the youngest of five children in a large, rambling three-story home, which was never without noise and activity, or friends of the children around the dinner table. In the summers the children would sleep on the roof-top captain's walk. In a countryside without hills they did their climbing by running up two flights of stairs to the captain's walk; then sliding down the roof to the conservatory, they would jump to the tin roof of the porch, crawl in the bedroom window, and start up the stairs again to run the circuit over and over. Easton may have been then, as it is now (6,800 people), the central town on the upper Shore, but it was still a small town and, like the rest of the Eastern Shore, its character reflected the steady life of the watermen and farmers it served.

But in the late 1920s Jim Rouse's secure world came unpinned. His father, who had periodically made and lost large sums of money as a canned-goods broker, suddenly lost everything. By 1929 both parents were seriously ill; they went into hospitals and Jim into a boarding house. His mother, who had suffered from heart trouble for a number of years, died that winter, and by August Jim's father was dead of cancer. In October the bank foreclosed on the Rouse home. Life for Jim Rouse suddenly became a struggle with poverty. His most treasured gift during those bleak years was a pair of track shoes that Dia and O'Donnell sent him so that he could still participate in school athletics. He later put himself through law school, parking cars during the day for $13.50 a week at the Saint Paul Garage in Baltimore and attending classes at night.

As Rouse's view of life was shaped by small town Easton, the Rouse Company would be shaped by his navy associations in World War II. The Rouse Company came together, years later, with the help of the officers Rouse joined at Pearl Harbor.

Rouse had been assigned to the staff of Adm. John Towers, the Billy Mitchell of naval aviation and then Commander of Air Forces, Pacific Fleet (COMAIRPAC), headquartered at Hawaii. When Towers was appointed COMAIRPAC, he immediately sought out for his staff a small, elite cadre of mostly Wall Street veterans of wealthy backgrounds. Jim Rouse was hardly of the proper bloodlines, but he knew the world of finance from his years with the Federal Housing Administration and his own mortgage banking company. He joined the Towers staff at Quarters 114, their duplex Shangri-La, on Ford Island in Pearl Harbor.

Many naval officers passed in and out of Quarters 114, but the original dozen men stayed together and transformed the two bungalows into one of the most sought-after watering holes in the Pacific. Assigned to the personnel group, Rouse saw to it that Quarters 114 had the finest food and service among any junior officer staff in the Pacific. They had servants, two Filipino stewards, their own mess, an ample liquor supply, and played nightly games of bridge and poker.

Rouse was to make many close friends at Quarters 114—Wing Pepper, now retired as chairman of the board of Scott Paper; and Harry Hollins, founder of the Institute for World Order; and some who would, after the war, become investors and directors in his company—Sam Neel, Lee Loomis, and August Belmont (now retired as chairman of the board of Dillon Reed). They found in Rouse an interesting combination of traits. Rouse surged with enthusiasm and optimism; he was somewhat naive, a bit straight-laced and overly trusting in others, thus easy, but good-natured, prey for practical jokes. But in other ways—particularly at the payday poker party, when Oakley Thorne, Towers's flag secretary, arrived brimming with cash—they discovered a Rouse more than their match. The game ran through the night and well into the next morning. Rouse paced himself, carefully measuring his drinks, the cards, and the players. After the others had stumbled off to bed, Rouse and Thorne would switch to gin rummy, at twenty-five cents a point. By sunrise Thorne would often find himself five hundred dollars down to Rouse. Rouse would feel guilty. "Guess how many chairs are in the next room—if you guess, we'll call it a hundred." During his entire tour of duty in Hawaii, Jim Rouse drew no navy pay; he lived entirely on his card winnings.

Any vote for the frumpiest-looking officer in Quarters 114 would have gone to Rouse, but his contemporaries then (as now) were intrigued with the contrast between Rouse's outward, unkempt appearance and the man they found behind it: a highly principled individual with a restless, independent, articulate, and far-ranging intellect.

A truly altruistic man. He has always stood out—principally for his exemplary values.

There's nothing phony about Rouse's liberalism, about any part of him, or about anything he says.

Jim is almost too good to be true. How could anybody be so motivated?

Rouse is an idealist.

Rouse, at first glance, appears extremely relaxed. He stands with a perceptible slouch. When he sits, he shoves his body deep into the chair until he is almost lying flat. Even his face is relaxed. But Rouse is, in the words of his first business partner Hunter Moss, a man of "inner intensity." His feet are seldom still. He digs the toes of his loafers together and snaps the edges of the soles across each other. If a low-lying table is near, he will work the welt of one shoe against it. One noon early in their shared careers, Moss, Rouse, and two business acquaintances were sitting on a padded bench in a Baltimore restaurant waiting for their table. Moss and the others gradually noticed that their foundation was vibrating. For a moment they might have considered this the overture to a mild earthquake, until they noticed Rouse at the end of the bench, deep in thought, his heel rapidly tapping the floor with sufficient force to jiggle four grown men.

During the fifteen years that he and Hunter Moss ran their small, but successful, mortgage banking business in Baltimore, Rouse pawed the ground, aspiring to nobler worlds. He hoped that in his later years he could free about 40 percent of his time from the business to deal with larger problems. These concerns drew him, for example, into the one-world movement (Rouse is presently chairman of the board of the Institute for World Order) and on a search for answers to the lifelessness he perceived in the American city. Rouse believed that people need not live in slums, and while serving on President Eisenhower's Advisory Commission on Housing Programs and Policies, he was the coauthor of a committee report, *No Slums in Ten Years*, proposing, optimistically, how to eradicate squalor. Rouse chaired the Greater Baltimore Committee, which redeveloped Baltimore's deteriorating downtown with Charles Center. He headed the American Council to Improve Our Neighborhoods (ACTION), which later merged with Urban America, Inc., and from which came, eventually, the Urban Coalition. "He always had fifteen balloons up in the air at the same time," Moss says. "He would run over and tap one up, then run over and tap another up." Sometimes he pops one.

The Moss–Rouse partnership began in 1939 as a two-man operation in one office with a secretary, but as it grew the two men agreed they needed a second secretary. A young woman answered their advertisement and seated herself in a chair before Jim Rouse's desk. He began to describe, with great elaboration, how fascinating she would find the mortgage banking business, and as he talked he fiddled with a long venetian blind cord that lay coiled behind his chair on the floor. Gradually—while maintaining the momentum of his job description—Rouse began to weave the cord through the arms of his chair, then through a few of the drawer handles, then back through the arms of the chair, crossing again to the remaining drawer handles in an elaborate web. The woman was unable to concentrate on what Rouse was saying, and could only respond perfunctorily as she gaped in disbelief at her intended employer, who was absentmindedly, but surely, imprisoning himself at his own desk. Finally, in complete exasperation she jumped to her feet, and blurting out, "I can't stand it!" she fled through the door.

By 1954 Rouse was ready to spread out. In addition to the mortgage banking business, he had organized a firm to advise developers on the economics of new shopping centers. Rouse decided he knew enough about shopping centers to build them himself, so he bought out Moss (an amicable parting), renamed the firm the James W. Rouse Company, and went into the development business. He began with his hometown of Easton, Maryland. Dia and O'Donnell Pascault were in the real estate business then. Both were convinced that Easton needed a shopping center and that Jim ought to build it. Rouse's initial reaction was that a shopping center in Easton was a ridiculous idea, but he reluctantly agreed with his sister to have his research firm look into it, fully expecting the data to support his instincts. But the evidence showed that Easton did need a shopping center. Rouse had another talk with his sister. We'll do it provided that it is not on the outskirts, he told her; find a site within walking distance of the center of town. The most appropriate land proved to be Lane's blacksmith shop and a few other contiguous parcels within two blocks of Easton's main intersection. It was here that Rouse built Talbottown, a small collection of retail shops along the architectural lines of the town's old buildings. It was Rouse's hope that Talbottown would help keep Easton from proliferating out along the highways, by drawing new development toward the heart of town. "There is no place where a shopping center is worse," Rouse says, "than on the outskirts of town. Walking down the streets, meeting people—that's what a small town is all about. When you lose that, you lose everything." But the highway along Easton's periphery now is like strip development anywhere in the country.

A year after Talbottown, Rouse's company opened the Harundale Mall at Glen Burnie, Maryland, and then Charlottetown in Charlotte, North Carolina, North Star Mall in San Antonio, Cherry Hill in New Jersey, and the residential community of Cross Keys in Baltimore. He was soon acknowledged as the innovator of the enclosed shopping mall and one of the country's foremost developers. As in the shopping malls, there is throughout the Rouse Company's new headquarters in Columbia a profusion of hanging pots of flowers, wandering Jew, spider plants, and great earthenware crocks that hold ficus trees that reach above the second floor. Rouse had seen the meager plantings of other shopping centers die from lack of care or from being improperly planted, badly selected, or poorly placed. When he built his first shopping centers, he was determined that the plantings thrive, so he hired the best gardener he knew, his sister Dia, and sent her out to supervise the landscaping of the malls. Dia was soon driving over 55,000 miles a year trying to keep ahead of the construction crews, who, when she arrived at a job, were usually about to run the plumbing and wiring right through the planting beds.

Years later, after innumerable battles with construction crews and labor unions, Dia informed her brother that hereafter she would be suspicious of the honesty of people until shown the contrary. Rouse's reaction was astonishment. "How can you say that?" he asked. "I would rather be gypped than have to believe that everyone is dishonest until proved otherwise."

That's his idealism for you, Dia says.

———— ✺ ————

Rouse's optimistic idealism had brought about Columbia, and it ultimately led him to Wye Island. (So optimistic is Rouse that when his parents died, he was stricken with guilt for feeling, despite his sorrow, he would become a better man because of his loss.) In both instances he wanted to create—in some ways, recreate—for others a place to live in which "life would be kinder, more humane, than the city." To Rouse that life was his boyhood Easton. "A small town is where people care for one another." That is what he intended to replicate for Columbia, when in 1963 he went before the Howard County commissioners to tell them that his firm had acquired roughly 10 percent of the county on which he planned to build a new town. He had in mind a number of small "villages," each with its own neighborhood school and shopping center. Although Columbia's center would be a large enclosed mall of restaurants and shops and a number of office buildings, the people would live in suburban clusters, surrounded by natural clearings and woods stitched throughout with bicycle paths.

Early on a recent summer evening Jim Rouse drove me around Columbia. He wheeled the station wagon around the curving streets to one of the villages, and stopped between two town houses. Behind them a wide meadow lifted from a narrow creek bed and spread to a line of woods beyond. "When Columbia is completed, better than 40 percent of the houses will be on open space," he said. "Not only will all the streams be preserved, but so will every little gully." Rouse turned up a street into Wilde Lake Village and drove along the edge of a Rouse-made lake. A sheet of water hissed over the sculptured dam. "They said it wouldn't fill for a year and a half. The lake filled up in sixty days." He slowed down in front of a frame house. It was one of the farmhouses within Columbia he had been unable to acquire. The woman who owned it would come out on the porch, gripping a rifle whenever she suspected a trespass. As Rouse turned the car around, a black man hollered to him and waved, and Rouse stopped to exchange pleasantries. As he pulled away from the curb, Rouse turned and said, "When he finally retired, our yardman of thirteen years moved into a house on Wilde Lake. The janitor, my day worker, and I all live in the same neighborhood."

Rouse has many lofty goals, but his highest is that people of different races and backgrounds can live together without conflict, and that poorer people should have decent housing. He has tried to make that vision a reality at Columbia, where about a fifth of the residents are black. Initially, he hoped to have 10 percent of Columbia's housing priced low enough for people of low and moderate incomes, but the collapse of the federal housing subsidy programs under the Nixon Administration and the erosion of Columbia's revenues by the 1974–75 recession have frustrated Rouse from meeting his goal. Though critics point out that the median annual income of Columbia's blacks is about $22,000, not exactly a sampling of the Negro poor, Columbia has ironically achieved one of its goals by being one of the few areas in the Baltimore suburbs where teenagers can find exciting things to do. However, some complain that not a few of Columbia's teen centers have become the exclusive domain of Baltimore's tough young blacks, and violence recently erupted there. Despite Columbia's problems—the worst being its financial problems aggravated by the recession—the new town only reinforced Rouse's belief that an open and fraternal society is possible.

But Rouse is also a businessman, and while making money—at which he has proved adept—is not one of the driving forces in his life, the success of his projects is. Rouse believes that "good development," as he defines it, will be successful, will turn a handsome profit for the Rouse Company, which by 1974 showed, in addition to Columbia, twenty-four shopping centers and nine office buildings completed and thirteen additional malls or expansions under development. He has tried to weave this blend of idealism and profit throughout the company and its projects. The effect, say his critics, is a form of positive thinking that can lead to presumptuousness and error. Others say that Rouse is simply a supersalesman: "He's so damned slick that he doesn't appear slick."

Initially, Rouse had been lukewarm about developing Wye Island. Other executives in the firm, such as Aubrey Gorman, had been more positive about it. What first haunted Rouse was whether this was what Queen Anne's County needed. The costs of the project, the high price of the island alone—$8,850,000—meant that only the well-to-do could afford to live there. Rouse was not comfortable with the prospect of building an enclave for the wealthy. During the debates within the company preceding the plan, one of Rouse's executives kept asking whether the development of Wye Island would serve a "redeeming social value."

During the presentation in the Centreville library, Robin Wood, the planning administrator, asked Rouse, "But could anybody afford to buy one of your estates or town houses unless he is pretty well off?" A troubled expression flickered across Rouse's face, and he momentarily looked down at his shoes. Then he looked up and quietly said no.

Rouse eventually decided to develop Wye Island for different reasons. The first was simply that he was convinced he could bring it off. It was the challenge, the horse that nobody could ride. Rouse is as addicted to that as any other developer. And the Rouse Company tends to believe it can take on the projects that no one else can. The second was that Rouse believed his project would be an economic plus for the county.

———— ⌘ ————

The yachtsman's vision of the upper Eastern Shore is a shoreline dripping with wealth—handsome waterfront homes and more than a few mansions surrounded by aging boxwoods. But for all its affluence (and in Talbot, particularly, it is considerable) life on the upper Shore does not come easily for many. For some, like the shanty-bound laborers for the seafood packers at Kent Narrows and the hundreds of other families who inhabit shacks around the region, living conditions are squalid. Nineteen percent of all the year-round houses in Queen Anne's County have only an outhouse for a toilet. Unemployment in Queen Anne's County, while much worse on the lower Shore, is nevertheless a chronic problem. Of those who are employed, half earn less than $8,200 a year, and one out of four workers makes less than $5,000 annually. There are few well-paying jobs in the county—many are barely at subsistence levels. Some natives sign aboard oyster boats in the hopes of making enough to buy their own workboats. But the seafood industry has always swung between feast and famine, and it is gradually assuming less and less economic influence on the Shore. Farming, as a source of employment, is even more discouraging.

As late as the 1940s much of Queen Anne's and Talbot counties were covered with vegetable crops and peach orchards, but today the vegetables and fruit trees are all but gone, and with them most of the field and cannery jobs. In line with national trends, farmers on the Shore use larger and more versatile equipment and fewer people. Frank Hardy can put up all his corn and soybeans on Wye Island with a power-steered, air-conditioned combine and a handful of men. With average interior farmland running around $1,100 an acre, combines costing up to $50,000, and fertilizer bills doubling within a single year, few sons of farmers can afford to follow in their fathers' occupation unless they inherit the family farm. And then only if they can pay the estate taxes. Those who do not follow the water or get into farming may find work at Freil's lumber yard, or scraping boats at the marinas, pumping gasoline at filling stations, repairing roads, or painting houses; many of the blacks face a weary future in the oyster-shucking stalls, as do the women splitting clams and picking crab meat at the shed tables.

But most young people drift away from the Shore in search of better times. So do many of their parents. Almost a third of the entire county labor force drives outside Queen Anne's County to their jobs, some to the towns of Chestertown, Easton, and Cambridge and others to the cities of Annapolis, Washington, and Baltimore across the bay bridge. The freight trains may soon be gone; the U.S. Department of Transportation has recommended discontinuing all freight lines into Queen Anne's and Talbot counties. There are few doctors, and no hospital, in the county. Twenty-five years after a group of Queen Anne's County citizens organized a physician's procurement committee, there are only five general practitioners in this county of almost 19,000 people. The committee has been re-formed for another search.

———— ✴ ————

The Rouse people were confident that their project would be an economic boost for the county, that the revenues produced would considerably exceed the demand for public facilities (which the company would build and donate to the county). They figured that the demands on schools would not be excessive, because about two-thirds of the expected buyers on Wye Island would probably be older, retired, or maintain only second homes. The property tax assessment for the island would jump astronomically. (Like most farmland in Maryland, Wye Island receives the preferential agricultural assessment that allows it to be taxed at well below its market value.) Development of the island would mean its assessment at fair market value, and that value would be significant, bringing into the county treasury well over $700,000 a year, almost sixty times the island's present tax revenue of less than $12,000. Rouse expected that over the estimated ten-year construction period, about 250 construction workers would be needed and that, ultimately, Wye Village and the estates would stimulate roughly 350 additional jobs for people in the county.

But the main reason that Jim Rouse decided to undertake the Wye Island project was to demonstrate how sensitive shoreline development could protect the environment and still accommodate the growing numbers of families flowing across the bay bridge in search of a place in the country. Rouse had little confidence that subdivision ordinances and green-and-yellow zoning maps would contain such growth. The only way to rid the Shore of shoddy development, he believed, would be to set the highest development standards possible in a living example on Wye Island. In the future, when the land hustlers showed up at the Centreville courthouse with hurriedly sketched subdivision plans, the planning commission would need only smile and politely say: "Go take a look at Wye Island. Unless you're willing to make that sort of investment, forget it."

First, the Rouse Company wanted to find out the environmental constraints of Wye Island and, particularly, its rivers. How much development could the island support, and what would be its impact on the surrounding area? Rouse had no blueprint, only an idea. He told his team of scientists and planners to make a complete examination of the island's limits and development alternatives. Raise the problems for us, Rouse told them. He was not too concerned, initially, about the costs of getting those answers, of creating the "right" plan. He wanted an exhaustive inquiry. It was. It ultimately cost the Rouse Company hundreds of thousands of dollars, and it was undertaken with only an option in hand, the zoning unchanged, and the prospects for amending the ordinance looking hazy at best.

The studies began soon after Rouse signed the option with Frank Hardy, in the summer of 1973.

CHAPTER 2

THE ROUSE PLAN EVOLVES

L ucien Brush and Charlie Flynn were beginning to wonder what would ever possess a sane family to pull up stakes and settle on the Eastern Shore. It was August 1973, and the only movement on the tepid Wye was the waves from their boat, as it hummed around Wye Island. The air was steamy and still, and hot. Brush, a professor of hydraulics and hydrology, and Flynn, Brush's graduate assistant, were enduring the first of several weeks around Wye Island, learning how its encompassing rivers move about within their natural containers. Brush and Flynn were part of a group of scientists from the Johns Hopkins University (though not associated with the university on this project) selected by the Rouse Company to investigate and document the physical and biological characteristics of the Wye River system. The scientists, headed by M. Gordon ("Reds") Wolman, the ebullient redheaded chairman of the university's department of geography and environmental engineering, were part of the larger team of

Rouse consultants and staff, who had been drawn together to determine the environmental constraints of Wye Island and its plan of development.

This much Brush and Flynn knew about the Wye River system: it is less a system of rivers than it is a small appendage of the Chesapeake Bay. Because so little fresh water actually runs off the land into it, the Wye River system is little more than a repository for the daily flushing movements of the bay's tides. Brush and Flynn wanted to know how well the Wye flushed itself, how well it could handle whatever wastes would flow into it from Rouse's proposed marina and island development. The best way to measure tidal velocity and circulation patterns would be to toss the dye rhodamine WT into the water and determine where it went and how fast it dispersed. But moving around the island, Brush and Flynn saw watermen in workboats checking their trotlines and crab pots, and they asked themselves, how would these burly fellows look upon strangers dumping a fluorescent, though harmless, blood-red dye into the river? Not too kindly, Brush and Flynn surmised. On a moonless night at about midnight, they shoved their skiff into the water and made the run from Wye Landing to a point off the island on the Wye East, shining a flashlight to keep from plowing into unseen banks. Here they dumped a fifty-gallon barrel of rhodamine WT into the river and ducked for home. Later they made another dye release in the narrows. In the ensuing days, between suffocating heat and drenching rain, Brush and Flynn encircled the island many times, taking water samples from marked positions and measuring the elapsed time of the dye's dilution. They found that the flushing ability of the Wye River system was moderately weak. But in the Wye Narrows, separating the Wye and the Wye East rivers, the water flushes hardly at all. Here the tidal velocities slow and finally stop. A marina on the narrows could create a stagnant mess.

While Lucien Brush and Charlie Flynn computed the Wye's hydraulics, Loren Jensen (then an associate professor of biology at Johns Hopkins and now head of an environmental consulting firm) probed what lived in it. Almost 3 percent of the total crab catch in Maryland—including some of its largest crabs—come out of the Wye. These waters are noted also for rich harvests of clams and oysters and excellent fishing. Jensen, however, would not settle for common knowledge. He wanted to confirm species variety and count their numbers, particularly the microscopic zooplankton, the plinth in the life column of most aquatic life. Jensen and his graduate students attached a fine-mesh net to a steel ring held between two runners and pulled it behind a motorboat, capturing zooplankton in the netted sled as it swooped and dipped like a roller coaster through the river depths. The zooplankton was preserved in formalin and taken back to the laboratory in Baltimore to be species typed and counted. Along the shallows fringing the island, the nursery grounds of juvenile fish, they deployed a fifty-foot beach seine. Through deeper waters they dragged a sixteen-foot otter trawl to snatch the adults—Atlantic menhaden, bay anchovy, oyster toadfish, variegated minnow, banded killifish, mummichog, striped killifish, killifish, mosquito fish, Atlantic needlefish, Atlantic silverside, silversides, white perch, pumpkinseed, yellow perch, bluefish, spot, goby, and hogchoker. Using a hydraulic clam dredge, Jensen and his students plucked up clams, oysters, crabs, worms, snails, and other animals from the river bottom. The Wye was full of aquatic life. The party slogged around the island and, assisted by aerial photographs, classified the wetlands. They analyzed water samples to assess the river nutrients—particulate and total phosphate, soluble orthophosphate, ammonia, nitrite and nitrate, organic nitrogen, soluble inorganic and organic carbon—and checked temperature, salinity, dissolved oxygen, and acidity.

Jensen would want to dip his nets and beakers in the rivers again during the spring spawning, but his preliminary examination confirmed that the Wye River system was biologically healthy. With one exception: the bacterial contamination of the clam beds. It took no laboratory analysis to establish this, for there were signs posted up and down the Wye and Wye East stating that clamming was forbidden by order of the Maryland Environmental Health Administration. Following tropical storm Agnes, in the spring of 1972, virtually all the clam beds in the Chesapeake Bay had been closed. Clams are highly susceptible to subtle changes in salinity and temperature, and when Agnes sent the Susquehanna into flood, many of them were killed by the massive rush of fresh water into the bay. Those not killed outright were badly weakened and as a result were unable to pump the tremendous quantities of water which clams need to absorb food and to purge themselves of pollutants. Health officials found clams almost gasping, their trunklike siphons hanging out by half a foot and their fecal coliform bacteria counts soaring. The hot summer further weakened the clams, and additional beds were closed on the Wye between Drum Point and the narrows.

Neither Jensen nor the state health officials, however, could pinpoint the precise source of the bacterial contamination, because when weakened by low salinity or warm water, clams can absorb high bacterial levels, even though the total number of bacteria in the water has not increased. Fecal coliform bacteria live in, and are exuded from, the intestinal tracts of warm-blooded animals such as humans, cattle, and geese. With most of the land in agriculture, relatively few people live within the small Wye watershed, leading the state to tentatively rule out inadequate septic fields as the cause of the contamination. Canada geese were on the suspect list, because at least thirty thousand winter as long as seven months on the Wye watershed. But the biggest source of bacterial contamination, the state concluded, was undoubtedly cattle. A single cow produces as much fecal coliform bacteria as do thirty humans. A handful of cows could dwarf the pollution from a village like Wye Mills. Cattle, pigs, and other livestock (one estate even has buffalo) graze along the Wye. A big herd of Aberdeen Angus is kept at Wye Plantation directly across from Wye Island. The Rouse team knew generally of the clam contamination, but they felt it should not foreclose the development of Wye Island since, at the outset, Rouse had instructed the project group that the entire island would have sewers, and no sewage effluent, no matter how purified, was to be discharged into the Wye. Rouse's decision, though praiseworthy from an environmental standpoint, was also pragmatic; he was almost certain that the county would not permit the discharge of effluent into the river.

Reds Wolman, the head of the scientific group, also hiked around Wye Island, exploring the coves by boat and in hipboots. He carried a hand core, and from time to time plunged it into the soft bottom of the coves, extracting cylinders full of sand and black mud. Later at the lab, these mud cores would help describe the rate and extent of sediment layering in the waters around the island. Wolman and his assistants pored over aerial photographs and charts of the Maryland Geological Survey to pinpoint where Wye Island's shoreline was eroding most, and where the sediment was being redeposited into spits and sand splays.

If left to its natural whims, the Chesapeake, in time, will consume the Eastern Shore. It does this particularly in the winter when, pushed by rough northwesters, it pounds waves against the exposed shoreline and grinds sheet ice into the soft banks. Like most waterfront owners on the Eastern Shore, Arthur Bryan, who lives on a farm farther up the river, has watched his acreage at Bordley Point on Wye Island gradually shrink and become the bottom of the Wye River. He has tried to hang onto his dwindling point by driving long timbers through piles of automobile tires stacked like poker chips along the shore. But in the half-dozen years since Bryan spindled the tires there, about eight feet of Bordley Point has washed away from behind his rubber revetments. But Wye Island, protected by Bennett Point, has not eroded nearly as severely as have more exposed places on the bay. In the century before World War II huge chunks of the western side of Kent Island— some more than a quarter-mile deep—washed into the Chesapeake. When Lord Baltimore took control of his province of Maryland in 1634, Poplar Island, a few miles below Kent Island, consisted of about 1,000 acres. By the 1950s, Poplar Island numbered only about 115 acres. Today it has been so carved away by the bay that its four small pieces total only 50 acres.

To the people who own eroding waterfront property, however, it is not the elements that they blame so much for their slumping embankments, but the rise in boat traffic. Roughly three-fourths of the power boats registered in Maryland are used on the Chesapeake: there were about 59,000 such boats in the state in 1962, ten years after the bay bridge was built; in 1975, almost 122,000, and the percentage that are trailered and brought over the bay bridge to the Eastern Shore is also increasing rapidly. Although almost everyone who lives on the waterfront has a pier jutting from his property and a boat tied to it, other boaters are viewed as an evil incarnate. Within those narrow creeks and waterways that do not face into the teeth of winter storms the wash from summer boats does cause erosion. Norman Briggs, a retired Exxon executive, who lives with his wife on Trippe Creek near Easton, has watched his shoreline recede about four feet over the past dozen years. He blames it on boats. Pointing over a raft of canvasback ducks which were feeding around the pier in front of his house, he said, "This summer there will be fifteen or twenty boats anchored overnight out there. . . more and more water-skiers and power boats. The wash is awful!" He stood on a long oak timber that braced a number of eight-foot-long two-by-sixes he had driven into the sand for his bulkhead. To cut expenses—timbered bulkheading can cost upwards from $150 a linear foot—Briggs had done the work himself, rigging an eighty-pound weight on a pulley to pile drive the timbers.

———— ∞ ————

Before the scientists had begun picking around Wye Island in the summer and fall of 1973, Bill Roberts, of Wallace, McHarg, Roberts and Todd (WMRT), had deployed his planning staff across the Eastern Shore to gather general data on traffic, schools, county ordinances, and growth trends in the region, in addition to making a general analysis of the island's environmental constraints. The climate was documented: forty inches of rain a year—wettest in August, driest in February. Prevailing winds from the northwest in the winter and from the south and southwest in the summer. Growing season: 232 days, with the first killing frost around November 17 and the last about March 30. Using maps of the U.S. Soil Conservation Service, they analyzed Wye Island's drainage patterns and soils—the Sassafras–Woodtown Association, the Matapeake Series, the Mattapex–Keyport Association, and the Elkton–Othello Association. Elevations and slopes, slight and subtle as they are on the Shore, were studied. In addition to identifying the trees on the island, Roberts's naturalists typed the various plants: yarrow, forking catchfly, field pennycress, moth mullein, Venus looking-glass, panic grass, and field brome.

From their research and the many conferences they had with the Rouse staff and the scientists, Roberts and his group began preparing the hundreds of maps, slides, and illustrations that they would use in assessing the environmentally strong and weak spots in Wye Island's capacity for development. There were maps showing prevailing winds, winter winds, and summer breezes. Others showed the island's geology, including six maps of each underground aquifer. One map on land elevations, two on slopes, one on scenic water features, and three on shoreline changes since 1847. Maps on soil types, soil permeability, erosion susceptibility, ponding suitability, and capacity for septic drainage. Maps showing vegetation, plant diversity, and the canopy height of trees. Maps showing aquatic life. Maps showing the best views down the river. Maps showing the worst views down the river. Maps showing no views down the river. All this, and more, was then put into the ultimate map—the "detailed synthesis," a WMRT specialty—where all of Wye Island's suitabilities and environmental limitations were overlaid in a display of arrows, legends, and colors that resembled the plans for the D day invasion of the Normandy beaches.

Ken McCord, a partner in the engineering firm of Whitman, Requardt and Associates, referred to his Wye Island assignment in this way: "We're strictly meat-and-potatoes people. We just want to know the facts." McCord has known Jim Rouse and worked on Rouse projects for many years. He is tall, exuberant, and plain spoken. Although both a registered planner and engineer, McCord has little patience with planning jargon. He does not refer to the small necks that define the coves on Wye Island as "enclosed anterooms off the central open space," as do Bill Roberts's planners and landscape architects. McCord calls them simply cornfields. The meat and potatoes that McCord's firm was undertaking to investigate and design for the Rouse Company included solid waste collection and disposal, electrical power, land clearing and grading, storm drain and sediment control, and, most important, water and sewer supply, and waste treatment and disposal. McCord is not inclined to view the physical requirements of development with the optimism that some engineers have for technological solutions. Ken McCord knew that the Rouse Company was willing to invest an extraordinary amount of money and professional talent in the planning, design, and construction of the Wye Island project. And he had great respect for Jim Rouse's infectious tenacity and his past success in moving seemingly immovable objects. But he wanted answers to his elementary questions: Was there enough water, Could the development be sewered without discharge into the Wye? McCord also knew that the Rouse Company had financial limits to its idealism.

McCord sent his associate Tom Shafer to the Eastern Shore for a reconnaissance. Shafer met with officials of the Delmarva Power and Light Company and was assured that ample power could be brought into Wye Island, and the Black Duck Refuse Removers, Incorporated, in Queenstown, would agree to haul away all the garbage and refuse.

So far, two potatoes.

But there was no company with which to contract for the water supply. The Rouse Company would have to drill for it. If a giant knife were sliced deep through the Chesapeake Bay and the Eastern Shore, and this piece of earthcake lifted out to be examined, it would reveal wedge-shaped layers of continental debris, which the bay-drowned Susquehanna had once carried toward the lowered sea, and deposits of shell, marl, clay, and sand laid down when the seas rose. Some of these underground layers, called *aquifers*, hold immense natural reservoirs of fresh water, which are fed by the percolation of rain water from the surface. The underground aquifers are the source of the Eastern Shore's water supply. Most of the farms and towns in Queen Anne's County have tapped into the hard beds of lime-cemented sand and soft greensand of the Aquia Formation, and Easton draws its water from the Magothy farther below. Louis Vlangas, McCord's geologist, took samples from some of these wells, including one on Wye Island, read through the research literature compiled by the Maryland Geological Survey, and talked with their experts. Was there enough water under Wye Island to support, say, ten thousand people? (The Rouse staff was using that figure for their preliminary assessment of development options.) The answer was yes, plenty of water. Getting it would require two wells drilled about five hundred feet into the Aquia and two more up to twelve hundred feet down into the Magothy. The iron content in the water, particularly from the Magothy, would require treatment, but Vlangas initially concluded that neither fluoride nor chloride from the aquifers would present any problem. Nor did he see evidence of saltwater intrusion into the aquifers. But he recommended that the Rouse Company first bore a test well into the Magothy and electrically log its performance to determine the water quality and aquifer's permeability.

One more potato.

But sewage treatment and disposal were complicated by the island's high water table and the poor permeability of much of its soils. The best solution was to collect the sewage on the island and pump it through a large pipe under the Wye River to Bennett Point, and then, by another pump, to send it twelve miles up to the interceptor sewer and across Kent Island to the county's regional sewage treatment plant near the bay bridge. Unfortunately, the interceptor sewer and the regional treatment plant existed only on paper. Queen Anne's County, like most of the Eastern Shore, has no regional sewage treatment system. Only a few of its towns, Centreville, Queenstown, and Millington, have any form of sewage treatment at all. When the toilets are flushed elsewhere in the county, the sewage simply runs out into backyard drainage fields. The county's sewage plan eventually called for a treatment plant on Kent Island, but that would be far into the future. The Rouse Company would have to build a sewage treatment plant for Wye Island, which later could be hooked into the county's interceptor. After considering the possibilities of building the plant above Bennett Point, and selecting a tentative site, the Rouse Company decided to locate it on Wye Island.

But this left few alternatives for the disposal of the sewage effluent. Shafer and Vlangas briefly considered the possibility of pumping it down into one of the saltwater aquifers, but the uncertainties of safely doing this without contaminating the freshwater aquifers left only one practical solution: spray irrigation. That meant constructing lagoons to store the effluent during periods when it could not be safely sprayed onto the land, such as when the ground would be frozen. Lou Vlangas made soil borings into Wye Island to locate the precise spots for watertight lagoons, and with a backhoe he dug twenty-one test pits around the island to determine which fields would best absorb the spray. The spray fields would have to be about 155 acres in size, with a surrounding buffer zone as large as 720 acres—a total of almost 800 acres for the spraying operation.

McCord and Shafer urged the Rouse Company to build, initially, a primary treatment system to screen the solids and aerate the effluent in the lagoons after killing the bacteria with chlorine; then, once the village was half completed, the plant could be upgraded to secondary treatment to draw out nutrients before spraying the effluent on the fields. Rouse agreed with the engineers that secondary treatment would adequately protect the Wye River system. But he wanted to take no chances with the volatile watermen, nor to leave any doubt about the purity of the water that would be sprayed onto the fields. As a political decision, Rouse told the engineers to design an advanced tertiary treatment facility—something that Wolman, for environmental reasons, had been urging all along. By the time the effluent reached the spray nozzles it would be clean enough to drink, far more pure than the present condition of the Wye. Rouse insisted that all costs of the sewage system, including the costs of later connecting into the county's future regional system, would be borne by the Rouse Company and the facilities donated to the county. But only if the county commissioners would amend their water and sewer plan to permit a tertiary plant on Wye Island.

The Wye Island bridge would have to be replaced. It was narrow and probably too rickety to carry cement trucks once construction began. Rouse's traffic survey showed that the county road connecting Wye Neck and the island to the main Shore highway would have to be widened and improved. Arthur Houghton wanted even more than that. He wanted both the bridge and the road moved almost a mile to the west. That would put the traffic a comfortable distance from Houghton's mansion on Wye Plantation. It would also conceal the crowds of chickenneckers, who pile onto the Wye Narrows bridge every weekend with their aluminum chairs and beer coolers to fish and crab, and who leave their trash for Houghton's farmhands to pick up every Monday morning. It would cost the Rouse Company a blue million just to relocate and improve the road but, for Arthur Houghton, Rouse was willing to dig deep.

Arthur Amory Houghton, Jr., controls the entire shoreline of Wye Neck that lies along the narrows across from Wye Island—and then some—and his influence is felt throughout the country. Houghton was founder and president of Steuben Glass. Since 1937, he had commuted on weekends between his New York town house and Wye Plantation. Recently, however, Houghton presented his town house to the United Nations to use as the secretary general's residence and moved to Wye Plantation. He was once the curator of rare books for the Library of Congress, and his library on the plantation, which contains among its treasures a first edition of Lewis Carroll's *Alice in Wonderland*, is considered one of the finest private rare-book libraries in the world. Houghton's philanthropy restored the seventeenth century Old Wye Church, and has put the Wye Mill, of like vintage, back into operating condition. He recently commissioned Elisabeth Gordon Chandler to sculpt a larger-than-life-size statue of Queen Anne (after whom the county was named) to be placed on the courthouse green in Centreville. Raised like laboratory specimens, the bull calves from his famous purebred Aberdeen Angus herd are weighed daily in their paddocks. The semen from his herd sires is sent to waiting cows around the world. Wye Plantation looks like one would imagine plantations are supposed to look. The pastures are bordered by rail fences and kept trimmed like fairways. Houghton keeps a small motor pool of self-propelled mowers. His staff drives about in black sedans equipped with two-way radios.

Rouse met often with Houghton and found him brimming with ideas about Wye Island. Houghton saw the possibility of an Aspen Institute of the East, an intellectual watering hole of rural elegance, where the World Bank might convene its board of directors or James Reston could lecture or maybe an occasional cabinet meeting could be held, should the president grow tired of looking out upon the Rose Garden. Houghton, like Rouse, wanted local people to use the village, so a movie theater and a bowling alley could be added as well. He hoped that the Rouse project would be an economic and cultural boost to a county that was losing its young people and had little to offer them. But the village would have to have a first-rate cheese shop; Arthur Houghton has to send clear into Easton for a good cheese. Wye Island should also have a fine equestrian facility, Houghton believed, where the Olympic equestrian team might practice.

Houghton also saw the development of Wye Island as a magnificent opportunity to demonstrate new seeds, crops, and cultivating practices on a scale even larger than that now used at Wye Institute, where Houghton runs a cooperative experimental farm with the University of Maryland, as well as art workshops, environmental symposia, and summer youth camps. Wye Island might once again be the experimental farming showpiece that it was two centuries earlier under the imaginative hand of John Beale Bordley. Michael Clarke of WMRT had been studying the Bordley era in detail, and he began urging that Bordley's principles be applied to the various open-space plans for Wye Island. Gradually, the Rouse staff grew excited over the prospect of linking up Wye Island's history by recreating Bordley's stewardship.

Judge John Beale Bordley owned the western half of Wye Island, from Dividing Creek down to what is now called Bordley Point, where he moved after retiring from the bench in Annapolis as one of the colony's last admiralty judges. Bordley believed the colonies could thumb their noses at England by growing and manufacturing everything they needed, and though few followed his example of self-sufficiency, Bordley left behind a rich legacy of experimentation.

When Bordley moved onto Wye Island in 1770, he dug his hands into its grayish clay soil and was appalled; the land had been drained of its nutrients by the earlier tobacco growers. Since the late 1620s, when the first European settlers landed on Kent Island, the Eastern Shore had been a vast land-clearing operation—clear the trees, plant tobacco, and then, when the soil was exhausted (usually after the fourth crop), cut more trees and plant a new field.

"The principal links in good farming," Bordley would later write in his *Essays and Notes on Husbandry and Rural Affairs*, "are due tillage, proper rotations of crops. . . and manures." But there were few cattle then on the Eastern Shore, so Bordley began shipping them in from England. Bordley was soon wintering on Wye Island a herd of 170 head, which he later described as "the old breed of the country, and of various colours, though mostly red, brown and brindled. About the year 1774, I began to mix this breed with a rather small but well-formed small-boned English breed. . . about the year 1785 these cows first had my fine bull, Horace. Horace and his sire had white hair on a yellowish skin, and their ears and noses were reddish brown." Bordley also brought 130 sheep onto Wye Island, each year shearing and selling the excess wool and realizing an annual profit of $109.50.

Livestock need salt, so Bordley drew water out of the Wye River, boiled it away in huge black kettles and spread the saline residue in the yard to dry. They also need fences. Bordley experimented. He first considered putting in post-and-rail fences, but after his men had gone a hundred yards, he ordered them to pull the posts. Bordley had been computing the costs of fencing half the island and maintaining them. He instructed his laborers to dig long ditches and plant them with thick impenetrable hedges.

With his once-famished soil now manured and replenished, Bordley planted the largest acreage in wheat and, to rest the soil, rotated crops. He threshed by pounding the wheat kernels from their husks under horses' hooves. Four teams of six horses each were hitched together in ranks around a circular track in a round room 135 feet across. A boy, seated on one of the lead horses, would walk the first rank over the bed of wheat. When that team had moved a fourth of the circumference, the next group of horses was started up, followed by the third and then the fourth, all two dozen plodding slowly around the room and crushing the wheat heads beneath. The horses would be walked six laps, trotted six, and walked again. After traveling eight or nine miles, they were rested, fed, and watered, the straw gathered up and the journey repeated. Bordley did not bother to send his grain up the Wye East River to Wye Mills to be ground but built a gristmill on the island and made his own flour. (Today on the Shore, wheat farmers and grain buyers wrangle over the amount of wild onions caught up by the combines; the buyers drop the price if the wheat contains too many onions. Considering some of the contaminants that Bordley's horses must have given his flour, the current onion dispute, by comparison, seems rather tame.)

John Bordley ran more than an experimental farm on Wye Island—he created a self-reliant community, albeit one dependent upon slaves. He built a kiln and manufactured his own bricks, pieces of which still stick out of the ground after Arthur Bryan's farm manager plows Bordley Point in the spring. In addition to wool from his sheep, Bordley grew flax and cotton, which was spun and woven into cloth for his field hands and colored with dyes made from his madder plants. He managed complete carpentry and blacksmith shops. He grew hemp and made his own rope, selling the unused portion to other rope makers on the Shore. In his eight-acre garden he grew every conceivable vegetable, spice, and herb. He raised bees and bucketed the honey. He shipped in Tokay grapes and cultured a vineyard. Bordley found that he could grow peaches and soon had a large orchard of fig, plum, pear, pomegranate, and even soft shell almond trees. From palmachristi he derived oil. Once his brick kiln began firing bricks, Bordley built a two-story granary, a double milk house, smokehouses, and a brewery (he also grew hops). Bordley preferred a chilled glass of imported Madeira, but he claims that the field hands loved his home-brewed beer. "It kept them in steady good heart," he wrote, "without any instance of such irregularity as rum commonly produces." Bordley even made a stab at distilling a treacherous brandy out of Irish potatoes, but failed to produce the liqueur of his expectations, a flavor "impregnated with the odor of violets and raspberries."

Both Bordley and his neighbor William Paca, who owned the other half of Wye Island and was one of Maryland's signers of the Declaration of Independence, helped supply George Washington's Continental Army. Paca shipped the army over a thousand flints for their muskets; he got them off the banks of the Wye, where English ships, which used flintstones for ballast, had been dumping them for over a century. Paca also melted down the ornamental lead from Wye Hall, the cavernous mansion he built for his son on Wye Island, and shipped it to the army to be molded into musketballs. They sent beef and wheat, too, although Bordley more than once impressed the supply ships for his personal use: the Governor's Council finally ordered troops to enforce the shipping orders should Bordley again interfere.

Judge Bordley once observed that the study of nature, natural history, and philosophy was the "most successful employment for a man in easy circumstances," and since those were his circumstances, that is how he would spend the day after his morning ride around the plantation to view all his enterprises. In addition to extensive writings on his philosophy of good farming, which were drawn from the experiments on Wye Island, Bordley applied his intellect to subjects as diverse as weights and measures, yellow fever, national credit, and moneys and coinage. After moving from Wye Island to Philadelphia in 1791, Bordley would establish the first agricultural society in Pennsylvania. His writings remain, but the rest of Bordley's imaginative enterprise on Wye Island has all but vanished—a few pieces of brick scattered about a cornfield. The mansion burned down in 1879. The two huge piers, granite blocks mortared in lead, have given in to the relentless waves of the Wye, and the subterranean passageway that once connected the mansion to the piers was filled in at least a century ago. Today when Arthur Bryan, the absentee owner of Bordley Point, steps off his front porch, he steps onto one of John Beal Bordley's millstones. The other stone is embedded in Bryan's lawn as part of the stone walk around his house.

———————

With Bordley's history in mind, the Rouse and WMRT planners began redesigning, on paper, Wye Island's landscape. To break up the dominance of corn and soybeans, they would bring wheat back to Wye Island, and millet, buckwheat, sorghum, oats, barley, rye, and cowpeas as well. Then, hay crops for pastured horses—probably a blend of timothy and clover—and grazing pastures, rather than just huge fields of corn, laid out near the houses and planted in bluegrass, brome, and fescues. They would grow orchards—apple, cherry, plum, pear, peach, quince, walnut, hickory, and filbert—and build an apple press in the village for homemade cider. Smaller orchards of fruit and nut trees for the wildlife would be planted along the hedgerows—which were now referred to as "upland ecotones." New woods would be planted, and portions of the fields abandoned to allow the natural forest succession: from sassafras, hackberry, mulberry, and dogwood to beech, oak, and hickory. The planners sketched in along the roadsides native shrubs like blueberry and huckleberry. Drawing on the Rouse Company's experience in Columbia, they set aside ten acres for a nursery to propagate the trees and shrubs. A vitaculturist advised that French hybrid and, possibly, vinifera grapes could be grown on the island, and from this vineyard a local wine produced. Vegetable gardens and berry batches would be encouraged. Wye Island would set a new pattern for commercial farming, using sewage sludge from the treatment plant rather than chemical fertilizers. To blunt eroding waves, plant salt marsh. To trap sediment runoff, seed the field edges in waving grass, which, if cut about six times a year, would spatter the tall green with black-eyed Susans, Queen Anne's lace, daisies, butter-and-eggs, butterfly weed, and goldenrod. New hedgerows—nonnatives like Russian olive, autumn olive, and Japanese barberry, but also native hawthornes and black cherry for their spring flowers and bird forage—would be planted along the property lines to break the wind and hold the soil.

On surveys of the island, Roberts's naturalists had seen redstarts, thrushes, warblers, thrashers, orioles, finches, bluebirds, bobwhite quail, towhees, and mourning doves, and they counted at least four osprey nests around the shoreline. They had also seen two families of red fox in a field, leopard and green frogs, snakes in the small drainage swales, and the flash of whitetail deer along the edge of the woods. They knew of the wintering waterfowl. The idea for a protected wildlife on the island now took shape.

Bill Sladen wanted Wye Island and, for that matter, most of the Eastern Shore to stay as it was. A physician and zoologist at the Johns Hopkins University and a world-renowned expert on swans (following their migrations north into the wild reaches of Alaska), Sladen believes in preserving huge areas of wildlife habitat from encroaching civilization. But he knew that Frank Hardy intended to dispose of Wye Island in one way or another, and Sladen did not want to see it covered with five hundred homes on five-acre lots. Sladen knew Rouse, having once done a year-long ecological study for him of Long Point. When Rouse asked him to design a wildlife refuge on Wye Island, Sladen saw the chance to do something imaginative in the middle of a development. Upon his survey of the island, Sladen was disappointed to find so few marshes, but he saw where new ponds could be dug as resting places for black and wood ducks. He knew that the geese would attract hunters, and he hoped that by diversifying the island's habitat, as the Rouse Company wanted to do, a greater variety of waterfowl might come in. Sladen did not want a government refuge; he considers the Fish and

Wildlife Service to be too hunting-oriented. He favored a private refuge, modeled along the lines of Sir Peter Scott's Wildlife Trust at Slimbridge, England, where hunting would be forbidden, where new marshes could be established, and where scientists could study how to propagate rare and endangered species of wildlife. So Sladen recommended that Rouse create a private wildlife trust on Wye Island. It would be donated to the Chesapeake Bay Foundation, run by a combination of the Johns Hopkins University, the foundation, and the Wye Island Association (to be made up of people who would move there), and affiliated with Peter Scott's Slimbridge.

———— ᘒ ————

This much was becoming clear to the Rouse people: the environmental constraints of developing Wye Island were entirely surmountable. Breaking up the large cornfields might mean a few less geese, but also it would mean less chemical fertilizer washing into the rivers. At Wolman's urging, many smaller storm drains, rather than a few larger ones, would be proliferated around the island, thereby helping prevent heavy storm runoff from tearing up the river bottom; the natural swales and high-grassed fields would also slow runoff velocity and trap sediment. By using a permeable pavement laid on beds of sand for roads and parking lots, much runoff would never reach the rivers. It would simply trickle underground, eventually to the aquifers below. No sewage effluent would be piped into the Wye, even though after tertiary treatment the water sprayed over the ground would be pure enough to drink—in fact, so devoid of impurities that it would taste flat. The Rouse team even considered banning all boat slips on Wye Island and buying the end of Bennett Point, where the Wye enters Eastern Bay, for the marina site. Do that and you can fold your tent, Rouse's market survey reported. Who will want to live on Wye Island if boats can't be kept there? Thus, using the detailed studies of water depths, wind exposures, sailing times, and relative flushing actions, the Rouse team decided to limit the number and types of boats and piers that people could have on the island and where they could go.

But as the planners began to brainstorm the development options for Wye Island, they hit a snag: Rouse, at this crucial stage, was inaccessible to all but a few of the senior company officials. No one knew exactly what he wanted for Wye Island, and, as it turned out, neither, at that time, did Rouse. In the absence of an agreed-upon concept, WMRT drew up seven development options, which were ultimately presented to Rouse in a series of maps and colored slides at a meeting attended by all the members of the project team. The first slide featured a literal interpretation of the option agreement: 2,500 houses scattered around the island, all the shoreline left in open space, and a town located smack in the center of the island. Next slide: a resort town, centered on Dividing Creek and surrounded by town houses and apartments; single-family neighborhoods were arranged along two golf courses, and all of Wye Hall Farm was to remain a farm. Next, the "shore-cluster" concept: mostly groupings of town houses in sixteen neighborhoods along the coves and rivers, with the island's interior left open. Following this came the "river hamlets" scheme, based on the shore-cluster concept, featuring only eight hamlets but doubled in size. Next, "Harbortown": one large village on Bigwood Cove, mostly made up of detached, single-family houses and a wildlife preserve on Drum Point, with Wye Hall farm left untouched. Finally, the "river/farm villages" concept: three town centers, three harbors, the west end of the island in farmland, a golf course on Wye Hall farm, mixed housing types around the coves—a density of 1,650 units. Roberts had searched through his files for illustrations of villages around the world—villas on the Mediterranean, mountain hamlets, Clare Village in Suffolk, England, and Sea Pines at Hilton Head Island. He had almost thirty different pictures of a "river hamlet" to show Rouse.

But Rouse was irritated. He did not want to talk about "urban images," housing configurations, or neighborhood locations. Long before looking at designs for Columbia, Rouse had wrestled with concepts of what Columbia should be and what it should stand for. That is where he was with respect to Wye Island after viewing the WMRT slides. Later in the fall and winter, as a specific plan began to take form, Rouse would immerse himself in such details. But that could wait. He asked each one around the conference table to say what he thought Wye Island should be. What follows is a loose paraphrase of the comments:

Wye Island is a perfect spot for a resort, like Hilton Head. Tennis, golf, the works.

Like it or not, only the gentry can afford Eastern Shore waterfront. Keep it low density, with grandiose estates.

You fry like an egg over there in the summer. Make this a place for second homes, where the cabinet types can spend the fall and spring. And we better have some swimming pools.

It could easily be for permanent homes. You know, for the guys who can commute into Washington and get away with being late to the office.

Should we even be over there?

We ought to replicate a true Eastern Shore village, like Saint Michaels. A place that the local watermen could use.

Oh, for Christsakes, find me an oysterman who's going to be able to afford to live in that village!

Within the Rouse Company this debate continued well into the fall of 1973, as one proposal after another for the development of Wye Island was produced, analyzed, argued over, and discarded. Three plans at a density of 1,700 houses, most located along the shoreline, were tried but rejected. The Rouse staff then put all the single-family houses in the center of the island, clustering the town houses and apartments in two small villages, one at each end of the island. No solution. Roberts and his staff sprinkled fourteen villages around the interior. Still no solution. "We need a stronger town center." A large village near the bridge with two smaller villages at each end was proposed and rejected, as was one for four smaller villages within a larger town. To illustrate what they did not want, a complete plan was drafted to show how the island would look if development followed the present zoning—one house for every five acres. This, too, was rejected.

Although the various plans were not discussed publicly, Doug Godine and Scott Ditch met with many organizations on the Shore to test local sentiment toward the development of Wye Island; most said, the fewer people on the island the better.

Over the ten-month period leading up to Rouse's presentation in the Centreville library on March 14, 1974, as many as twenty-three alternative development plans were drawn up, considered, and discarded by the Rouse Company. These were not twenty-three hurried sketches, with black blobs to suggest a village and casual circles for the open space. WMRT had rendered no less than 118 different topographic layouts of Wye Island, on at least four different scales. Each plan was laid out exactly as it would be constructed. For some, as many as six architectural layouts were made, not only for the overall plan, but in varying scales for each of the neighborhoods, so that the drawings included every shrub, tree, walkway, tennis court, and driveway, practically everything but the dogs in the backyards. McCord's draftsmen plotted on graph paper the exact elevations, grades, and profiles of Wye Island and from these made drawings and cost estimates for the installation of the roads and sewer and water lines for each plan. But throughout this period of examining alternatives and testing them against costs, potential market, and what they guessed the county might accept, indecision prevailed.

The costs of each proposal varied, but some items, involving tremendous amounts of money, stayed about the same for all plans. The company would spend about $2 million to landscape the island and reforest some of the fields. It would cost $1 million each to stabilize the shore erosion, to build a new bridge, and to improve and relocate out of sight of Houghton's mansion the road leading to the island. Another $1 million would be needed for the eventual hookup to the county sewer system. Sewage treatment (including screening and grinding, metering, activated sludge, clarifier, chlorinization, advanced treatment, controls and operations building, dewatering, site work, electrical work, and thirty acres of lagoons) would come to another million. The cost of wells, $180,000; water treatment (iron removal, wet well, sedimentation, pumps and controls, chlorination, a building to house the treatment plant, and mechanical, electrical, and site work), $650,000; an elevated water tank, $125,000; sewage pumping stations, $250,000; water transmission and sewer interceptor mains, $3.5 million, plus; $400,000 for the force mains; and $2.5 million for the main road and connector roads on the island. Each neighborhood would have to have storm drains and channels, curbs and gutters, sediment control, fire hydrants, blowoffs, valves, borings, manholes, sewer collection and force mains, pumping stations, house connections, and land clearing. Rouse has a thing about signs: they should be pleasant and unobtrusive. Fifty thousand dollars for a sign program. Costs for the boundary survey, $30,000; the solid waste survey, $10,000; a fire and police station, $500,000; paths and trails, $750,000; $500,000 each for twenty acres of lakes and ponds and the building of a tennis club; and $1.5 million for a golf course and club house. The Rouse Company planned to invest about $100 million in Wye Island.

By October 1973, when Rouse had originally hoped to have a proposal for the county, the optimistic mood at the Rouse Company had darkened to frustration. One plan after another had been researched, designed, and costed through the company's economic model, but Doug Godine's team had been unable to agree on how many houses should be built on Wye Island. "We were strictly confused," said Godine. "Just like an artist who has gone through canvas after canvas, we had gone through plan after plan, and none of us were really satisfied with where we were." The economics of the project—staggering costs, however configured the ultimate plan—loudly argued for a population range of seven to ten thousand. The simple arithmetics meant that as the numbers of people on the island decreased, the costs of each housing lot rose almost proportionately. But after each trip across the bay bridge, Ditch and Godine returned with the same gloomy faces. They met with different groups, but the message rarely varied: "If you Rouse people can do what you are saying you can do, that will be a good thing, but you're just saying words. People ruin the rivers, people cause problems. People don't have to come over to the Shore. Not if we resist."

Godine walked into Rouse's office. If we go to the county with a proposal for more than a thousand homes on the island, he said, we don't stand a snowball's chance in hell of getting approval. Besides, it would be wrong, Godine argued, to put so many people on that island. Wye Village, or villages, would be bigger than Centreville and Queenstown. Instead of being down in the boondocks (as the natives considered it to be), Wye Island would soon dominate the county. Few of Rouse's fellow directors were bullish on Wye Island. Auggie Belmont, who had retired from Dillon Reed to a waterfront home less than a mile from the island, wished that Wye Island could remain as it was. He liked to gather oysters off the submerged granite piles that were once Judge Bordley's piers and fish for perch and rockfish there. When first approached by Frank Hardy about Wye Island eight years before, Sam Neel had been somewhat pessimistic about local acceptance of Wye Island's development, and he still was. Some corporate officers thought that the Rouse Company couldn't fit its social goals into the recreation and second-home markets. Others felt Jim Rouse was being a bit too generous with the company's resources on a project that the county officials had never shown any interest in approving. Rouse was caught in a riptide—the harder he swam, the less progress he made.

Rouse himself had been initially skeptical about developing Wye Island, but Rouse, once enthused, is not easily dissuaded. He thrives on obstacles, almost enjoys demonstrating how they can be overcome. "Life should work," he says. "Most things that one would rationally want to do are doable." Columbia had only reinforced his optimism. When he first announced that he was going to build Columbia, Rouse had plenty of skeptics, including the Howard County commissioners, recently elected to reverse the easy rezoning of the past. Rouse went all over the county meeting with any group of citizens he could. When Rouse finally went before the commissioners for zoning approval to build a new city where there were only dairy farms, Columbia got a unanimous vote. To get your way in life, you have to fight, his sister Dia used to tell him. Rouse would always correct her: "You don't have to fight. You have to persuade." To Rouse, the overriding importance of Columbia is that it exists: almost 35,000 residents, over 400 businesses, 80 industries, 17,000 jobs, and a sixth village recently opened. Columbia has a fundamental community ethos, says Rouse: a high sense of the *possible*. If we could pull off Columbia, we could sure as hell do Wye Island.

But Rouse agreed with Doug Godine that the plans to date were all out of scale with the Eastern Shore and would have to be scrapped. Earlier in the summer he had looked through about two hundred WMRT slides of several Shore villages. The photographs had captured the characteristics of the villages—the steeply sloping roofs, backyard boatsheds, narrow alleys, and short blocks. And the challenge of recreating one of these villages now captured Rouse. But time was running out. It was now October 1973. The Hardys' option had to be exercised before the end of June 1974. Rouse had precious little time to test public acceptance of his plan before having to pay $8,850,000 for 2,500 acres of cornfields and woods or drop the project entirely. Godine's team went back into high gear to design yet another Wye Island, this time with a single village. Bigwood Cove was finally chosen as the site for the harbor and village. It offered the best views down the Wye, and by having the deepest water of any of the coves, it would not require as much dredging. Laying U.S. Geological Survey topographic maps of Saint Michaels and Oxford over Bigwood Cove, the planners replicated the street layout of the old villages. The number of houses for the island was finally settled upon: 890 units. The economic model fairly groaned, and Rouse was finally forced to make a significant compromise. With the overall density so low, the costs of building sewers to the scattered 184 estates were prohibitive. That meant a septic system for each estate.

In late February 1974, Lou Vlangas, the geologist, hired a backhoe operator and dug test pits on Wye Island to determine the soil's percolation capacity for handling septic leaching fields. Vlangas was alarmed to find that a number of the places along the west end of Wye Island failed to meet the percolation requirements; water poured into the test holes at the bottom of these pits did not drain out in the time required by state law. But by now the public meeting for unveiling the plan had been delayed over five months, and Rouse wanted no further delays. If necessary, he concluded, some of the estates could be enlarged to avoid placing houses near the poorly drained soil. Some of the house sites undoubtedly would have to be relocated. Conceivably, the western end of the island might have to be connected to the central sewage system for the village. But Rouse insisted on keeping the March 14, 1974, date in Centreville for presenting the plan.

———— ✿ ————

Jim Rouse has taken about fifteen minutes to sketch the outlines of the proposal to the people crowded about the scale model in the library. He asks Doug Godine to present some of the details to the planning commission members. There will be no apartments or condominiums in Wye Village, Godine says, only town houses. "Yeh, *row* houses like in downtown Baltimore!" snorts a voice. Others join in the snorting. Many on the Shore are immigrants from Baltimore, or, as they say, *escaped* from Baltimore. Much of the older part of Baltimore is lined with block after block of marble-stooped, brick row houses mostly housing Baltimore's poorer and working-class families. And many black people. To many who live on the Eastern Shore, particularly white people, all the fright and ugliness of living in a city are symbolized by Baltimore's endless row houses. Arthur Bryan says of Columbia, "Rouse took that beautiful county and cut it up into row houses. They call 'em town houses, but they're nothing but *row houses*!" A row house in Baltimore is apparently of a different species than those on the Shore. In the middle of Centreville, a block and a half from the library, there is a short street that flanks one side of the courthouse and green. It is called Lawyers' Row, one of the town's most historic streets, the prestige address for most of the county's lawyers. The law offices are converted row houses.

"What is the need for a two-hundred-room inn?" demands Mrs. Eugene du Pont III, her eyes flashing. The du Ponts live across the Wye East River from Wye Island on a big estate. Of all the wealthy landowners along the Talbot side of the Wye East whom Rouse had consulted about the plan that winter, the du Ponts had been the most intransigent. Before Rouse can answer, Bobby Price, the attorney for the planning commission, asks Rouse how he plans to judge public opinion. Price's ancestors owned a large part of Wye Island during the nineteenth century. "Will there be a vote?" he asks.

"Obviously not," says Rouse. "If the people of Queen Anne's County say they don't want our development, we'll never file for rezoning. And someone else will develop the island in the conventional manner. If the people indicate they like the proposal, then we'll proceed." In every public utterance Rouse has made this statement. He firmly believes that, in time, he can meet with enough people in the county, show them the advantages of his plan and answer enough objections to eventually defuse the opposition. That is how Rouse laid the groundwork for Columbia, and how he thinks he can overcome resistance to his plan for Wye Island. When people on the Shore realize that the alternative to clustered Wye Island is incremental sprawl, they will come around, Rouse believes.

A number of questions are thrown at Rouse at once, most of them from Mrs. du Pont, all suggesting that no matter how he tries to limit the number of boat slips, Wye Village will attract plenty of outside yachts, and all those toilets will just be flushed into the river. There will be places at the docks connected to the village sewage system to empty the holding tanks, he replies. There are groans.

Mike Thompson's face is reddening in anger. Mike Thompson is a native from Grasonville. For the past six years he has lived along the Wye River above Bennett Point. From his boat dock, he can look directly across the river to Grapevine Cove—into the edge of Rouse's Wye Village. Thompson is on the planning commission. He leans across the scale model and jabs a finger at Bigwood Cove. "Why did you pick *this* place for a village?" he demands. "The oyster bars below and above it are now polluted and closed! They've got signs up!" The library room is suddenly and uncomfortably still. The du Ponts smile at each other. The members of the planning commission exchange looks. Then they stare at Rouse. Jim Rouse says nothing. He has nothing to say. Incredibly, after ten months of exhaustive studies, water sampling, zooplankton counting, slide taking and mapmaking, no one has informed the company president of the health administration's latest shellfish bed closure. Rouse turns to Bill Roberts, who says mildly, "We picked it because this area has the best flushing action."

Thompson's voice is now strident. "Then all you're going to do is add more pollution and simply flush it around!" There are more snorts and barely suppressed chuckling. Roberts does not say that his staff has put together a fifty-five-page boating analysis of the island and the rivers. Rouse does not refer to the twenty-three alternative plans and the months of study and effort that had led to the scale model before them. Jim Rouse decides it is futile to argue now. He turns to Mike Thompson and agrees that, if after further testing and analysis, it is confirmed that the pollution potential from a marina at Bigwood Cove is as serious as Thompson says, the village site will be relocated.

Suddenly, after only forty-five minutes, the meeting is over. Robin Wood announces that the commission has a prior commitment at the courthouse, and the commission members promptly walk out a side door. Jim Rouse says he will be more than glad to remain and answer anybody's questions, but there are virtually no more questions. Eugene du Pont smoothes the lapels of his pinstripe suit. The room soon empties. Rouse slouches in a folding chair, his hands in his pockets, legs stretched out and his feet switching in frustration. His attorneys disassemble the scale model and begin carting it up the street to the storefront that the Rouse Company is opening. Rouse is let down, drained by the apathy. The Rouse Company has mailed to every resident a richly illustrated brochure of the Wye Island plan; a local editorial said more about the costs of postage than it did of the merits of the plan. Rouse cannot understand why there has been so little reaction to his plan for Wye Island. "If you exclude the county officials, the press, and my staff, there probably have been fewer than ten to twelve natives show up," he says.

Doug Godine walks over and says, "I counted only two today."

CHAPTER 3

THE NATIVES

Sam Whitby didn't attend the Rouse meeting in Centreville. He went clamming that day, which he does on most days when the weather is not too rough and his helper shows up. But Sam probably wouldn't have gone to the meeting in any event; like most natives he doesn't care much for public meetings. He knew Rouse was a native of the Shore, but that didn't weigh much in Sam's thinking since Rouse was no longer "from" the Shore. Sam did not share Rouse's vision for Wye Island. Sam grew up there. Although little about Wye Island suggested a romantic past or an imaginative future to him—he remembered a lonely, sometimes stark existence on a remote island—Sam was uncomfortable about the way the upper Shore was changing, about the city people who were moving in, and about the Rouse project.

If Columbia were Rouse's plan for tomorrow, the natives wanted no part of it, particularly on Wye Island. So they stayed away from the library meeting that March day in Centreville. They climbed up on their tractors or boarded their workboats, and left Rouse at the library to face an audience composed largely of people who had recently retired or moved to the Shore from the cities across the bay.

It was 4:30 on a cold spring morning. The stars were still bright. Along the waterfront the homes of the retired and the mansions of the gentry were dark and quiet, but in the villages and across the flat farmland the lights were coming on in the clapboard houses of the watermen and the farmers. No Mercedes on the highway at this hour—only big semitrailers roaring down from Wilmington and pickups heading for the landings, taking the watermen to their boats.

Sam Whitby did not disturb his wife Lillian when he got up on this morning, so instead of the eggs and scrapple she ordinarily fries him, he fixed himself a bowl of cornflakes and ate alone. Sam Whitby is seventy-two years old. He has worked with his hands, his back, and his legs all his life— farming, carpentering, oystering, crabbing, and clamming—and his short but stocky physique shows it. On the water he wears a green cloth cap, and except for his forehead, which is white where the cap sits, Sam's wide, solid face is the reddish brown of men who work out-of-doors. When I drove into his yard, just outside of Wye Mills, to join him for a day of clamming, Sam was leaning against the cab of his pickup, smoking a cigarette.

"Morning," he said. "We'll stop in Grasonville and pick up something for lunch." We got into the truck and drove up the highway toward Kent Island. Sam stubbed out his cigarette in an ashtray crowded with wads of chewing gum. He unwrapped a fresh stick of Wrigley Spearmint and curled it into his mouth.

We drove past Hickory Ridge Estates, one of many farms Frank Hardy is subdividing, over the headwaters of the Wye River, and into the little hamlet of Grasonville. Sam parked the truck in front of Leonard Smith's lighted grocery store and went inside.

"Morning, Cap'n Sam," said the rotund county commissioner, who stood behind the counter reading the weekly paper. Just about every waterman on the Shore who owns a boat is called "Cap'n," usually, in conjunction with his first name, rarely his last.

"Morning, Leonard." Sam walked directly to a shelf, snapped off two bananas from a bunch, picked up a couple of packaged apple strudels, laid them on the counter, and asked Smith for a carton of Kents.

"D'ja see this, Cap'n Sam?" asked a bleary-eyed man in a hooded sweat-shirt. The man, another waterman, stood at the door next to a stack of county newspapers. Whitby walked over and stooped down to read an article headlined: "Clamming Industry Faces New Problem." Additional clam beds were being closed because the heavy spring runoff from the Susque-hanna had lowered salinity in the upper bay.

"Hmpf," said Sam, shaking his head slowly. He looked at his watch and squinted into the dark through the front window. "Lister should've been here ten minutes ago," he said. He walked out into the black morning, slightly scowling. Bill Lister has been Sam's helper on the water for ten years (Lillian refers to Lister as "Sam's man"). Lister works hard once aboard the boat, but he is something less than dependable about showing up. Because of his age, Sam no longer takes his boat out in rough weather. Consequent-ly, his modest income depends on getting over the clam beds on every good day. Yet on more mornings than Sam likes to remember, he has stumbled out of bed when the sky is like frozen ink, driven up to Lister's house in Grasonville, and banged on the door and thrown gravel at the window in vain. Lister drinks only beer, but he has been known to work through a case with dispatch.

Sam walked back inside Smith's store. A few other watermen had arrived, bought beer and provisions for their boats, and gone. Sam glanced at his watch again. By state law the clammers have to quit by noon every day. Sam was losing precious time. He walked back outside.

"There he is," Sam said, and without another word he got into the pick-up and started the engine. Out of the blackness and into Sam's headlights shuffled Bill Lister, his eyes red and swimming. Two front teeth were miss-ing; the rest were the color of broomstraw. His clothes looked like he had slept in them—often. He had not shaved for at least a week, and his face was red and raw. After we met, I asked Lister where he lived.

"Down at the end of that lane. On the marsh," he pointed vaguely down a dark street. Bill Lister smiled. "I'm the poorest man in the county," he laughed, as he squeezed into the cab. Sam gunned the pickup onto the highway.

Sam drove across the drawbridge at Kent Narrows onto Kent Island, then swung southward down the island, through the cluster of ramshackle houses called Dominion, and pulled up to a small wharf on Little Creek at the top of Eastern Bay. Dawn was beginning to show the silhouettes of the workboats. We went aboard the *Lillian A.*, and Sam started up the engine. It uttered the unmistakable guttural of a Buick, which is what it was—a 1971, 350 horsepower, Buick automobile engine. The engine and transmission had cost Sam $1,500. Sam switched on a light in the cramped cabin and propped open the cabin window with a ball peen hammer. He seated himself on a stool behind the wheel and rubbed the condensation off the windshield with a paper towel. The ledge under the window had wads of chewed gum stuck to it. As Lister cast off the lines, Sam backed the gargling *Lillian A.* into the creek and began his run into Eastern Bay. It was cold. A gusty northeast wind smacked the flood tide coming up the bay, creating large swells, but by running with the wind the *Lillian A.* rode smoothly through the chop. Sam said that the return trip would be rough. Parson Island passed to the left and then tiny Bodkin Island on the right. Four miles to the east, Bennett Point was only dimly visible—and obscured behind it, Wye Island.

Whitby and Lister debated where they should look for clams. Lister insisted that Captain Emerson Tarr would know. Sam said nothing. Lister reached overhead, switched on a two-way radio, and tried to call up Tarr. Another clam boat responded on Tarr's behalf. "Cap'n Emerson's got no market today," the voice said. Most of the watermen have standing arrangements with particular seafood packing houses to purchase their clams or oysters, but when demand drops, the packers adjust by telling some of the watermen they will not buy that day, or for a number of days. A clammer or oysterman thus turned away is said to have "no market."

Sam Whitby squinted through the sprayed cabin window at Tilghman Point, a couple of miles off the bow and said, "I think it might be good over to Claiborne. I think I see some boats. Let's just slide over there." Suddenly, the radio erupted into the sound of an imitated police siren: "Aaaaaaaaah-hhhhhHHHHHHHHAAAAAaaaaaaaa! Miles River Bridge! Red Dog, Red Dog is on theee way!" A fellow clammer up the Miles River was warning other watermen of the position of a marine police boat. Most clam boats on the Chesapeake Bay are now equipped with two-way radios. Sam initially resisted buying a two-way radio, until his only son and grandson were saved by having one. His son was halfway across the Chesapeake Bay one morning in his oyster boat, when the wind velocity suddenly jumped from fifteen to forty-five knots, boiling seas as high as a house. "Electrolysis had eaten away the nails in the bottom, and when she came up on a big wave, the water just sucked the bottom out and she went down in a few minutes," Sam said. "Both my son and grandson were in the water, and that water was thirty-five degrees. Another five minutes, and they would have been dead." Before hitting the water, Sam's boy radioed a distress, and a nearby boat heard the call in time to save them. Soon after, Sam drove down to a store and bought a Johnson Messenger 350 and installed it in the *Lillian A.* None of the watermen communicate by their call letters. "If they don't call us by name, we don't know who they're talking about," Sam said. When Sam painted his boat, he painted over the radio call number. He has written the number on a scrap of paper, which he keeps in his pocket.

The *Lillian A.* was forty feet long, rather wide, descending gradually to the water from bow to stern. She weighed seven tons. Sam built her about ten years ago out of yellow pine, which he seasoned and then treated with a mixture of pine oil and turpentine. He had to bend the boards by hand. Sam said he'd never build another forty-foot boat. The keel weighed over eight hundred pounds, and Sam and his son had to hoist it into place from the bed of a truck. The small cabin was forward to provide ample work space on the wide deck. The Buick engine that powered the *Lillian A.* was secured amidships for stability. Bolted down near it was a smaller Chevrolet engine, which ran the hydraulic motor at the stern, which in turn drove the conveyor belt on the clam rig. Sam built the clam rig, as he did everything else on the boat, except the engines. Sam's clam rig, like others on the bay, consisted of a steel-mesh conveyor belt, which was supported by a long rectangular metal frame and hung alongside the boat from davits at each end.

By 7:00 a.m., Sam reached the mouth of Eastern Bay, where it opens into Chesapeake Bay, about a quarter of a mile off Wade's Point below Claiborne. He cut the engine and the *Lillian A.* began to pitch and roll in the freshening wind. With a power winch, Sam eased the head of the clam rig onto the bottom of the bay, and then cranked up the rear of the rig by hand so that Lister, at the stern, could stoop over and reach the conveyor belt. He kicked on a hydraulic pump, which sucked water out of the bay on one side of the boat through a six-inch hose laid over the deck, and forced it down through a four-inch pressure hose and out eight nozzles affixed to the nose of the rig. As the water suddenly boiled around the bow, Sam switched on the conveyor belt, which rattled like a box full of rocks thrown on a tin shed. He restarted the Buick engine and from a wheel at the stern slowly guided the boat along an unseen path of buried clams. The jets of water blasted into the sand, dislodged clams, shells, and rocks and threw them onto the conveyor belt. Lister stood at the stern, where the rear of the conveyor belt emerged at an angle from the water, and with the quick hands of a crap game croupier, he plucked off the legal-size clams, tossing them into a bushel basket between his legs. The combined roar from the Buick, Chevrolet, pump, hydraulic motor, and clattering steel belt was deafening. Within half an hour, Lister had heaped high the first basket with plump soft-shell clams. He stacked it aside on the washboard, and began filling the second. Sam turned the boat around and started back alongside the first course.

———————cᐱɔ———————

Sam Whitby grew up on the western end of Wye Island, on the "Bryan place," as he calls it. His father moved the Whitby family (five sons and two daughters) there in 1907, when Sam was five years old, to "tenant," or manage the place for the Bryans, who had long since moved up the Wye to another farm. Sam stayed there until well into his teens, moved away to Easton for a few years, and then returned to marry and tenant-farm around the island until 1936, when the Stewarts bought much of the island and evicted the Whitbys and most of the others.

One afternoon, a few weeks after Rouse unveiled his plans, Sam revisited Wye Island. He drove down the dirt road to where it ends at a solitary wooden gate, got out, and leaned on the gate to look beyond at the farm of his childhood. Like the rest of Wye Island, the Bryan place looked little like Sam had remembered it as a youngster. Although the Whitbys had been poor, Sam's father had run a well-kept farm. From the barnyard five fields of corn and wheat had fanned out like slices of plum pie, each field neatly bordered by a Bordley ditch row. His mother, who died there when Sam was thirteen, used to scrub the pine floors of the farmhouse with lye every week so they gleamed like a brass buckle. The house had been surrounded by English walnut and white heart cherry trees, and an orchard of plum, apples, and peaches had spread from the house to the river.

Now all that Sam could see was the abandoned farmhouse and an empty field. The orchard was gone, as were all the barns, sheds, and ditch rows. The field had been plowed from the river banks to within a few feet of the sagging porch. Pieces of plywood had been nailed across the doors and windows. The only sign of life was a woodchuck that whistled from a burrow next to the porch. Sam remembered how he and his older brother had clipped, or "brambled," the low hedgerow of Osage orange that once flanked the gate on which he leaned; he also remembered their orange fights with the drupes that grew from it. Sam stepped back from the gate and looked up at the line of trees on both sides that towered more than fifty feet above where the hedgerow had been. The trunks of some, streaked with orange, were over two feet thick and covered with huge thorns.

In the distance the woodchuck scooted across the empty lane. During the spring rains the wagon wheels would mire so deeply in the mud that Sam would stack bricks in the ruts of the lane and tread them down with the mule. "There used to be a powerful lot of bricks down here then," Sam said. Judge John Beale Bordley's bricks. Rains would seal the soil into a hard crust that a plow had difficulty breaking up, so Sam would hook behind it a subsoiler, with teeth that burrowed more than a foot below the plowshare. Occasionally, the mules would be thrown on their rumps when the subsoiler caught the edge of Bordley's buried tunnel.

Sam turned and looked back up the island. A mammoth yellow pine once grew near the gate, and on Sundays the three tenant farmers living at that end of Wye Island would meet under its shade and trade the week's news. The tree is now gone. Sam Whitby fixed me with a look that expressed neither nostalgia nor regret. He simply wanted to make a point. "That's all there was then," he said.

Sam's most vivid memory of Wye Island was its remoteness. The winters then were more prolonged, the snow deeper, and the rivers and coves were often frozen solid all winter. The Bryan place at Bordley Point was the farm farthest from the narrows bridge and the villages beyond. Today you can easily drive from the gate at the Bryan place to Centreville in half an hour. But in the early 1900s, the journey by horse and wagon over boneshaking, rutted dirt roads took half a day. As a boy, Sam Whitby rarely got off the island more than twice a year.

The first trip would be around Thanksgiving. Like most tenant farmers, Sam's father was allowed a winter "lay-in"—flour, meal, meat, kerosene, and a few other essentials. This meant a trip to Centreville. Sam and his father would catch their turkeys, load them in a high-sided wagon, and cover the bed with fence gates to keep the turkeys from flying off. To anchor everything down, they would heave sacks of corn and wheat on top of the gates. At three the next morning they would hitch the horses to the wagon and set off for Centreville. The horses had to be walked slowly to keep the wagon from pitching in the ruts and distributing the turkeys all over Wye Neck. They would arrive at Centreville late in the morning, and, after selling the turkeys, return by way of Wye Mills, where there has been a mill operating since 1672. (The Whitbys live only a short distance from the mill, which is now operated by students from a nearby college.) After the corn and wheat were milled, Sam and his father would head back toward Wye Island, stopping briefly at the tiny settlement of Carmichael to complete the lay-in with beans, coffee, sugar, and kerosene. The sun would be low in the sky when the wagon crossed the narrows and started down the island. At the barns they unharnessed the horses in the dark.

If it had not snowed heavily and made the roads impassable, Sam might get the chance to ride into Queenstown around Christmas when his father brought the harness or saddle in for repair.

Other than those two trips, Sam Whitby saw little of the world beyond Wye Island. And Wye Island was a pretty lonely place. There were few boys Sam's age to play with, and they were far apart: Howard Melvin lived a two-mile walk up the road; Ralph Whaley, another mile beyond, at Bigwood Cove; and Grason Chance was clear on the other end of the island. To this day, Ralph Whaley regrets never having learned to play team sports, like football or baseball—there were never enough boys to field a team. When all the children gathered at the schoolhouse, they rarely numbered more than eighteen, ranging from six year olds up to men in their early twenties who were still in the seventh grade.

Sam drove up to the center of the island and stopped next to a jungle of undergrowth. "Everybody walked then," he said, "and it was a right good walk, too. You think of a child these days walking three miles to school through mud and snow. . . that's what we had to do." He looked down the road in the direction of the Bryan place. "You hear about the good old days," Sam continued. "Well, it wasn't good old times to me. It was rough going."

Sam stepped across a ditch of black water, and waded waist deep into an almost impenetrable thicket of woods, honeysuckle, and catbrier. He was searching for evidence of his former schoolhouse. The vines snarled about his knees and shins. No structure was visible, but Sam kept turning to get his bearings from the road. Finally, triumphantly, he planted his foot on a mound of honeysuckle and began rocking a rotted timber that lay under it. Sam said that it was one of the sills—now all that was left of the school-house. He described a small building about sixty feet in length and forty feet in width, which had been the scene of six of the seven years of his only formal education. "I know it had four windows on each side and two in the front. . . best I can remember it," Sam said. "I had a teacher once reached right out the window and grabbed me by the hair. I was fighting with an-other boy. That's how she parted us. She reached right out the window and grabbed me by the hair."

Because Wye Island was so inaccessible, the teacher boarded at a nearby farm. A small shed behind the school housed the coal for the stove. Ralph Whaley got five dollars for firing the stove all winter, building the fire on Monday mornings, and banking it each night until Friday, when he allowed it to die out. There was no pump at the schoolhouse, a delight to the two boys detailed from classes for the daily bucket run. They took their time finding a distant pump and returning with the bucket, slung between them, porter fashion, in the notch of a stick. The children gathered around and drank from a tin cup, which was passed from mouth to mouth.

Sam Whitby stood motionless for a while, his hands in his pants pockets, staring into the green gloom of what was once the schoolhouse yard. "Our blackboard was boards about six inches wide, painted black, and you had to write across those cracks, some of them a half-inch wide," he said. He had been up since 4:00 a.m., working his clam boat in the bay, and though he'd had a short afternoon nap in his reclining chair, as he always does after a day on the water, Sam's heavy shoulders slumped from fatigue.

The isolation of Wye Island drew the farm families together to cut wood, slaughter, and put up the corn and wheat. For over two centuries Wye Is-land had been cultivated and grazed, but trees were abundant, and still are. White, red, and chestnut oak stood along the river bluffs and, with willow and black and postoak, dominated the woods. Scattered among them were huge loblolly pines, and mockernut and pignut hickories, tulip poplar, beech, and sweetgum. In the fall each farmer cut and hauled enough logs into his yard for about twenty cords. Then the farmers gathered together at one farm, and using a thirty-inch circular saw powered by a belt from the flywheel of a Peerless steam engine, they cut, split, and stacked the firewood—moving down the island to each farm in turn. (The Bryan place had few trees, so the Bryans shipped logs across the Wye to the Whitby's on the Drum Point ferry. The ferry was a rectangular scow. A rope, stretched across the Wye between two locust poles, was passed through two rollers on the scow, which was powered by the ferry operator and the passengers pulling the rope as they walked from fore to aft.)

At butchering time the women prepared the meals—fried chicken, shortbread, and tea—while the men killed the hogs, drew and hung the carcasses, and cleaned the intestines for making sausage. The next day the men skinned and butchered, and the women stuffed the sausage, made scrapple, and baked the pork, sealing it in jars with the grease.

Summer was, as the natives say, "thrashin'" season. It commenced usually after the Fourth of July, at the farm where the threshing machine had been left at the end of the last season. All the farmers and their sons met there and began putting up the wheat. The wheat was cut and bound into sheaves by a horsedrawn binder, stacked tentlike into larger shocks, and carried by wagons to the thresher. The thresher was powered by the same steam engine used to cut the wood. It held six barrels of water, and needed continuous refilling. That was the job of the young boys, who drove a team and wagon from the threshing field to the nearest pump and back all day in the heat. They threw grain bags over the tops of the barrels to keep the water from sloshing out. After finishing the first farm, the crews moved on to the next.

At each landing around the island, mules hauled the grain wagons into the river and the men shoveled the wheat aboard pungey boats, to be taken out to large schooners anchored in deeper water. Wye Island has some of the best soil in the county, and, in the early 1900s, when most other Queen Anne's County farmers were averaging about twenty-two bushels of wheat per acre, the farmers on Wye Island were harvesting thirty-five to forty.

Each family made their own butter then, and sold the excess. The Whitbys sold theirs in Saint Michaels (fifteen cents a pound). They hailed a nearby crab boat (in the summer) or an oyster bateau (in the winter), and sailed, or rowed, the six miles out the Wye and across the mouth of the Miles River.

Sam Whitby holds no idyllic illusions about the good old days. "You know," he said, "this was a pretty dreary place down here. No electric, no mail, no nothing."

————— ✧ —————

Life on Wye Island was not all work, however. Like the other boys, Sam Whitby spent most of his summer Sundays exploring its coves in search of anything out of the ordinary. In the shallows around the crumbling remnants of Judge Bordley's stone wharves, Sam would slowly wade, with the patience of a heron, his eyes and hands searching for arrowheads, or Indian darts as he called them. When Sam looked up his friends, he rarely wasted time walking around the lengthy shoreline but simply swam across the coves to the next farm. He and his companion would catch a calf and try to ride it. "We was just looking for sport," Sam said. Ralph Whaley and Howard Melvin each had bicycles, which they often took across the Wye on the rope ferry and spent the day riding up and down the Bennett Point road. But, like Sam Whitby, Ralph Whaley used his spare time mostly exploring Wye Island. He called it plundering. "My brother would sit in the house and read, but I was always a 'plunderin'," he said, "poking around in those coves. . . see a tree, and see if I couldn't climb it." In the winter, the coves became skating ponds.

If it were not plundering, it was fighting. Living so far down at the remote end of a remote island made Sam quite shy, and when he first started school, he stuck close to his sister. He gradually overcame it, however. Those boys who worked around him on Wye Island, or wrestled with him at the schoolhouse, considered Sam Whitby a powerful adversary for his size. As a young man, he could press a hundred-pound keg of nails over his head twenty times before putting it down. A break in the threshing schedule was as good a time to fight as any. Two pairs of boxing gloves hung from the Peerless steam engine, and during a lull between loads, Sam boxed anybody who would put on the other gloves. "We didn't have much else to do, so we'd fight," he said. "I used to come home every day tore up. If you box without taping your hands, specially your thumb, it tears them up."

Sam's happiest times occurred during the gunning season. In the early 1900s, when Sam was growing up on Wye Island, the Chesapeake Bay swarmed each fall and winter with great rafts of canvasback, redhead, and scaup ducks which gorged themselves on the wild celery beds of the Susquehanna and other rivers. Flocks of them often traded back and forth over the Whitby's house. "Boy, there were ducks here then," Sam said. "Not many geese, though. And you could hunt anywhere you wanted, 'cause there was no one around. My father would tell my brother, 'You better take your gun out'—he had an eight-gauge—and with one shot he often brought down ten ducks. That's how thick duck was then."

Sam first started hunting with his father's muzzle-loading, ten-gauge shotgun. It had forty-two-inch barrels, and when he placed the butt of the gun on the ground and held it out to one side, the muzzle was over his head. Occasionally Sam loaded it with too much powder. After he pulled the trigger and the air cleared of the gray, acrid smoke, Sam picked himself up and found unburned black powder all over the snow. When the bruising recoil finally cracked the stock, Sam simply wired it together and kept hunting.

The early part of the century was the waning days of the market hunters. At night, crouched behind their skiff-mounted, multibarreled battery guns, they would quietly slip up on a large raft of sleeping ducks, slap the water with a paddle to bring their heads up, and then touch off the small cannon with a deafening roar. The more successful market gunners often picked up more than ninety ducks after one of these blasts. Sam Whitby was not a market gunner. He did not sell the ducks he shot; he gave them to his family or to friends. He had none of the expensive equipment of the gentlemen shooters from New York, Philadelphia, and Baltimore, but, like them and most natives, Sam Whitby loved to hunt ducks. He shot his wherever he could find them, and he was not above crawling through the snow to a cove and taking ducks at rest on the water if he could. He once got seventeen redheads with one shot this way,

Gunning was a passion with Sam Whitby, which even poverty could not abate. Sam was resourceful. To save enough for his first shotgun, he husked corn every day after school. On the next lay-in trip off Wye Island, he stopped at the Carmichael store and bought a single-barrel, twelve-gauge Iver Johnson. It cost Sam six dollars, which he considered a small fortune. But he needed shotgun shells, and shells were expensive—forty-five cents a box in Saint Michaels. So he sold oysters. During the summer when the oysters spawned and were flaccid, Sam would strip off his clothes and wade around in the Wye shallows, plucking oysters off the bottom. He piled them in a skiff, rowed around to a small cove behind the farmhouse, and dumped them in. In the fall Sam would push the skiff into the cove and with hand tongs bring up the oysters, fat and edible, selling them to a buy-boat out of Baltimore. He got to Saint Michaels by arranging for a ride with one of the dozens of oyster boat captains who used the cove as shelter during the night.

The Whitbys did a brisk business with the watermen who overnighted in their cove, selling them potatoes, meat, eggs, clabber, and butter. Few of the boats carried any cash, so the Whitbys took fish in payment and let the crews sleep in their hayloft. At night the aroma of frying potatoes in dozens of cramped galleys would lift from the cove and spread across the farm.

Tilting back his head and closing his eyes, Sam said, "I can almost smell it now."

One crisp October day Sam returned again to Wye Island, and when he got out of the car at the Bryan gate he inadvertently slammed the door. The scene exploded, as thousands of Canada geese thundered into the air from the Wye River and spread out across the sky, in vast undulating waves. Sam practically had to shout over the roar of wings and honking. "That's a cove over there where those geese are getting up," he said. He turned to watch a dozen wheel overhead and set their wings for the unseen cove. Over 600,000 geese now winter on the bay, the first flights in from Canada arriving in late September. They stay as late as early May. The geese have increased on the Shore by hundreds of thousands since the mechanical corn picker was introduced; it leaves enough unshelled ears on the ground for a winter's goose feast.

Sam has about given up waterfowl gunning. One reason is his age. The other more determinative one is that there are few places left on the Shore for him to hunt. Farmers and city retirees now lease out their fields and shorelines for such high gunning fees ($3,000 to $5,000 a farm for the season is not uncommon, and the price keeps going up) that people like Sam can no longer afford it. He shook his head and said, "I didn't even buy a license this year."

Sam had known Lillian Porter most of his life. The Porter farm was just across the river on Bennett Point, and Lillian spent the summers on Wye Island with Sam's sister. As a boy, Sam did little to establish the best foundation for his later marriage to Lillian—once he locked her in the meat house and yelled that the place was full of rats. But within a year after the Whitbys moved from the Bryan place to farm "the Mansfield place" (where Sam's father had been born) near the Wye Island bridge, Lillian and Sam were married. They moved a short distance to a small house at the head of Granary Creek, which they rented from a Mr. Robinson for three dollars a month. (As elsewhere in the country, the names of particular places on the Shore are as numerous as there are namers. Sam Whitby refers to it as the Mansfield place, because a man by the name of Mansfield once tenanted it; other natives call it the Tim Bishop farm because Bishop once owned it. It is now only a sweet corn field of larger Wye Hall Farm and called, by its last owners, the bridge field, because it is nearest the Wye Island bridge.)

From the cove behind the Robinson place, Sam set out in a log canoe to hand-tong for oysters.

————— ᴄⅣᴐ —————

For years the log canoe had been one of the mainstays of the Chesapeake oyster fleet, used by those who could not afford to buy or build the larger skipjacks and bugeyes (which under sail gathered oysters by means of a large dredge rake and twine bag which was pulled by cable across the oyster beds and handwinched aboard). The log canoe was made of three dug-out logs joined together, pointed at both ends, and was without a deck. On its mast was a leg-of-mutton sail and a jib. Like almost all forms of bay oystering, hand-tonging was, and still is, backbreaking work. The hand tongs consist of two wide metal rakes attached to long wooden shafts (anywhere from eight to twenty-six feet long—the short ones are called "nippers"), connected like scissors by a pin of dogwood. The oysterman stands on the narrow washboard, lowers the tongs to the bottom, and scissors them back and forth until he feels he has "a good lick" of oysters in the rakes. Then he pulls the tongs, clamped together, straight up, hand-over-hand, pivots them on his thigh, swings the rakes over a culling board in the center of the boat, and dumps his catch. He repeats this procedure until the culling board is full, then he lays the tongs across the boat and separates the legal-size oysters from the catch with a small hammer. He does this in the winter usually alone, encumbered in hipboots, when the washboard may be covered with ice, and when a fall into the water paralyzes the body—a fast way to drown. People from the cities who do not—and never would—work this way for a living consider the handtongers a romantic breed.

Hand-tonging today is done the way it has always been done on the bay, except now the workboats are powered by gasoline engines. The bugeyes are few and the log canoes are rapidly disappearing. The rest of the oyster fleet consists largely of "patent-tong" boats—which must remain stationary, and drop and lift their rakes hydraulically from booms—and about thirty skipjacks, which by law are still required to operate under sail, except for two days of the week when they can use their pushboats for power. The skipjacks of the Chesapeake are the last commercial sailing boats in the United States.

———————— ◌◊◌ ————————

In the summers Sam laid out a half-mile-long trotline for crabs. Along it he set up to six hundred salted-eel baits, then, sailing by in the canoe, he pulled the trotline over a pole and scooped off the feeding crabs with a long-handled net. Lillian supplemented their meager income by raising and selling turkeys and sewing dresses, and, later, by upholstering divans and overstuffed chairs.

Entertainment on Wye Island, as elsewhere on the Shore in those days, was unfrilled and infrequent. When the ferry operator, Mr. Sylvester, purchased the first phonograph, people walked in from all corners of the island to hear music on cylinders. When Sam and Lillian finally saved enough to buy their first radio—a battery powered Scott–Atwater—friends would join them on Saturday night to listen over earphones to the fading plunks of Nashville banjos. No one thought much of walking three or four miles across snow-covered fields to play cards or visit. They didn't play for money, because no one had any. In Sam's father's day, no one had matches so they "borrowed fire" from each other during the winter; if you cut a social visit too short, your host might say, "Must've just come to borrow fire, huh?" Twice a year families gathered at the schoolhouse for box socials, and in the fall for an oyster roast. On some Sundays the minister from the Carmichael church rode down to Wye Island and held afternoon services at the schoolhouse. Once in a great while Sam and Lillian rowed their skiff down to Saint Michaels to see a silent movie.

After two years on the water, Sam decided to learn the carpenter's trade. As he put it, "If you haven't got any education, laboring is hard, but carpentry paid right good, and I learned fast."

The Whitbys moved to Easton for five years, until the Depression dried up carpentry jobs, then they returned to the island to manage the Whaley farm at Bigwood Cove. Sam built a thirty-three-foot deadrise boat, from which he oystered and crabbed as slack periods in the farm work permitted. Later, he would build most of the post-and-rail fences that now frame Arthur Houghton's Wye Plantation. When the bridge to Wye Island began to deteriorate, Sam helped drive new pilings and rebuild it. It was a drawbridge, tended and hand-operated by Lillian's father, who sat in a tiny shack, which Sam also built, waiting to crank open the draw at the shriek of a boat whistle.

Like most natives, Sam and Lillian Whitby have worked hard physically—building, fixing, mending, scraping, pulling, lifting. They have known no other life. During the Depression, the Centreville merchants who displayed the Blue Eagle of the National Recovery Administration ran an ad in the local paper imploring people to start buying: "It must come from you, you patriotic women who hold the purse strings. . . . To increase wages. . . to increase employment. . . business must increase and that means buying must increase. Prices are going up!. . . The quicker you buy the more you save, and the more you help things right now." But Sam and Lillian could be of little help, for they barely made enough to subsist.

Everything was saved, nothing was wasted, and the Whitbys still live that way. Behind the chicken shed in their yard at Wye Mills there is a smaller shed in which Sam stores his goose decoys—the same shed Sam built about fifty years ago to shelter Lillian's father at the Wye drawbridge. Rather than build a new one, Sam hauled it off Wye Island when he left in the mid-1930s. A few years later, W. H. Stilwell, a wealthy New Yorker, bought the old Paca estate on the east end of Wye Island and proceeded to build a replica of the fire-gutted Paca mansion, Wye Hall. Stilwell told Sam that he could have the old Mansfield house if he could move it off the island. "It was the first time I ever had an eleven-room house and no place to put it," Sam said of the house where his father was born. He scavenged everything off it that was still usable. When he built the frame house where he and Lillian now live, Sam used much of the flooring and windows from the Mansfield house. The two pine doors into their bathroom are from Wye Island.

———— ⌀ ————

The conveyor belt suddenly stopped rattling. Sam quickly shut off the Chevrolet engine and winched the clam rig out of the water. A black timber was wedged in the frame between the belt and the nozzles. As the boat rolled in the seas, Lister climbed out on the rig and, after much effort, kicked it free. Rocks as big as a mailbox, bricks, and other objects often jam the conveyor belt, and Lister has to crawl out over the water to knock them off with the ball peen hammer. Sam said that a friend of his once plucked a pirate pistol off the belt; Sam's ashtray in the cabin is the broken neck from a clay jug that came up one day with the clams. Sam pointed to a large, white-columned home on the shore, called Mary's Delight, and told me that forty-five years ago he had helped the owner rebuild it. (In the 1830s it had been one of the earliest hospitals on the Eastern Shore.) The old kitchen had contained some interesting bricks, Sam said, which were dumped offshore when the "new" kitchen was built. Later in the morning Sam pulled a piece of brick off the conveyor belt and showed me part of the kitchen floor from Mary's Delight.

The wind was growing stronger and colder. Bill Lister asked if Sam would spell him at the conveyor belt. Lister walked over to the roaring Buick muffler and warmed his hands, and Sam began culling clams. He had maneuvered the *Lillian A.* over a full bed, and the clams were coming up the belt as thick as cookies on a sheet. Culling clams requires a strong back, good eyes, and fast hands. Sam still has the hands of a boxer.

The cutting wind streaked Sam's face with tears. He wore blue sneakers, lead-gray cotton work pants, a faded cotton shirt, two worn cardigan sweaters—one gray, one maroon—and his green cap. When Lister returned to the belt, Sam stepped over to the stern wheel and pulled a sweatshirt over his head. Eleven bushel baskets of clams now lined the washboards. The limit was fifteen.

All morning Sam stood at the wheel, shifting his weight with the bucking of the waves, and boring the *Lillian A.* along an unseen path, as neatly as one mows a rectangular lawn. His only reference points were the distant featureless shore and a couple of far-apart buoys that marked the landward line within which clams could not be taken. But in that pitching water, so far from any landmarks, Sam knew where he was at all times. Occasionally he would reach out and lay his hand upon the frame of the clam rig and by feel know whether he was pushing through sand, passing over a rock pile, or drifting uselessly over a deep hole.

The skipjack captains feel their way across the invisible oyster beds all winter in somewhat the same manner. As they sail back and forth through the crowded oyster fleet, they reach out and grasp the twanging cable and by touch discern whether and how fast the dredge bag is filling with oysters.

Like people anywhere who have lived in one place their entire lives, the natives of the Eastern Shore have an intimate sense of their surroundings. Most of the families who lived on Wye Island during the early 1900s still live today within a seven-mile radius of the Wye Island bridge: the Bryans, the Warners, the Melvins, the Shawns, the Whaleys, the Chances, the Bishops, the McClyments, and the Whitbys. Although it has been forty years, and more for some, since they left Wye Island, they remember the island as if their departure had only been yesterday. Howard Melvin says he can draw a map of every field on Wye Island. Sam Whitby says that if you took him blindfolded to any place along its river banks he could tell instantly where he was. "I *know* this island," he says. One afternoon as Sam drove me along the island road, I noticed a large white house in the distance, but because of the flat terrain and the trees along the banks I could not tell whether the house was on the island or across the river. I asked Sam if he knew whose house it was. He continued driving, without so much as a sidelong glance toward the river, and said, "If you see one, it would be across the water. That would be George Bryan's place. A lawyer lives there now."

Sam Whitby has the same instinctive sense about the weather. Only a careless waterman is indifferent to it, for being able to sense an approaching storm can mean the difference between coming home with a catch or drowning in the bay. Even the most experienced and cautious watermen have been caught unexpectedly on the water by a sudden squall. Out of necessity they have learned to read the subtle shifts in the wind and the patterns of the clouds. One summer afternoon in the Whitbys' small parlor Lillian showed me some old photographs of Wye Island. The day had been hot, and although the windows were open, the air was still and oppressive. Sam, barefoot, lay back in his reclining chair and relaxed after a day of clamming. Toward sundown a wind came up. Sam wriggled his toes in satisfaction as the breezes swept through the house. When I was leaving, Sam walked me outside. On the porch stoop he stopped and said, "We may not go out tomorrow. It looks like we'll get a northwester." The wind was coming out of the south. I asked Sam how he could tell that it would so shift and grow too strong by morning. He looked over his shoulder at a few wispy clouds now barely visible across the peach-colored southern sky. "I don't know," he said, "I can just tell, I guess. A fellow who depends on the weather can look at the sky and know more than the weatherman." The weather report for the following day had forecast a quiet bay. That night Sam did not bother to set his clock. By morning a northwest wind with twenty-six-knot gusts was driving the bay into an angry froth, and Sam Whitby slept late.

———————∽∾———————

By noon the washboards of the *Lillian A.* were crammed with 15½ heaping bushels of clams. About the extra half bushel, Sam said, "By the time we get back, these clams will have settled right low. We use the extras to top 'em off." The run back up Eastern Bay to Dominion was as jolting as Sam had predicted. The *Lillian A.* wallowed through the troughs and smacked against every wave. Bill Lister bent over the baskets, now safely on the deck, and inserted a small white card in the rim of each, which read:

SOFT-SHELL CLAMS
RIVER: EAST. BAY
AREA: WADE'S PT.
DATE: 5–2–74

By the time we reached Little Creek, the clams had drained and settled well into the baskets, and the extra half-bushel contained exactly enough clams to refill each basket. At the wharf we loaded the heavy clam baskets into the bed of the pickup and drove down to Kent Narrows, where Sam sold them for $10 a bushel to United Seafood. Out of this $150, Sam paid Lister $30; his gas cost about $35. Two months later the Chevrolet engine burned out, and Sam had to buy a replacement.

It was midafternoon when Sam drove up to his house. He had been on his feet in that bucking, rolling boat for almost eight hours. The bananas and pastries had only kept the edge off hunger. Sam and Lillian ate lunch, and then he took off his shoes and lay back in his reclining chair for a nap. About four o'clock he would go out to their large vegetable garden at the rear of the house and chip away the weeds with a hoe. After an early supper and a little television, Sam would fall asleep. At four-thirty the next morning he would get up and do it all over again.

———— ✺ ————

I dropped by the Whitby's house many times during my drives about the Shore. We would start talking about the weather and what kind of a day it had been.

"How was it clamming today?"

"It was beautiful on the water today, just beautiful!" Sam would say.

One afternoon I noticed that the field around the house had been freshly plowed. I asked Sam what he was planting. It was barley, but not to harvest, he told me. Later he would plow it under and add grass or clover. Sam has an honest esthetic sense. "I just like the look of green grass in the field," he said.

Books about the Eastern Shore usually extol its beauty. Sam Whitby says, with no embarrassment, that the Eastern Shore "is too damned flat." (The highest place on the Eastern Shore is less than a hundred feet above sea level—once the beach ridge of the sea when, thousands of years ago, the ocean was higher than it is now.) He likes the rolling pastures of western Maryland better, and his eyes widen and brighten when he talks about the mountains of Montana. The farthest the Whitbys have traveled from the Wye is Niagara Falls, but among their photographs and postcards is a scene of cattle grazing somewhere in the Rockies. Sam turned the photograph over slowly. "I love to look at the mountains," he said. He lay back in his reclining chair and looked longingly at the ceiling. "I always wanted to go to Montana." A friend of his once invited Sam to drive with him to the beach on the Atlantic. "Why do you want to go to the ocean?" asked Sam. "There's no scenery there, not down to the ocean."

Sam and Lillian have not lived on the water since they left Wye Island. Although living on the water seems to be the chief attraction to people interested in moving to the Eastern Shore, Sam and Lillian nevertheless have had many attractive offers to buy their house and six acres. Why hadn't he accepted one of those offers, I asked. "I'm too old to move," he answered.

"If I was younger, I might have moved to Florida. Oh, I could use the money, don't get me wrong. I'm getting too old to face the water every day. This summer I'll probably sell my boat. One man was standing on the pier with eight thousand dollars in his hand, shaking it at me! It was pretty hard to resist." (A year later Sam did sell the *Lillian A.* But he didn't quit the water. He bought some yellow pine and spent the winter building a thirty-footer. Last summer he fought his way through the weekend crabbers, and this winter he has been hand-tonging.)

Sam's Chesapeake Bay retriever Duke uncurled from under the yew hedge and trotted over to Sam's side. "I know almost everyone within a twenty-mile radius," he said. "You just can't get that in the city. They're too busy to get to know each other and be friends." One day Sam drove a relative from Baltimore around the Wye, and from time to time Sam pulled over to stop and chat with someone he knew. "Sam, you've met more people in just this one morning than I know in the whole city of Baltimore, where I've lived all my life," his relative told him. On the Whitbys' fiftieth anniversary, their son Philip had a surprise party at the Carmichael church; over 150 people showed up. Sam rubbed Duke's ears and looked at the house he had built with his own hands. Lillian walked over to the coop to let out the chickens. Sam turned and looked me full in the face. He said, "I've been here all my life. Where would I go?"

———— ᐧᑫ ————

Herbert Goldstein, who grew up on the "Western Shore" (as people on the Eastern Shore refer to the rest of the state of Maryland) and who moved to Centreville in 1938, feels he is part of the Shore now—as much as any outsider is ever fully accepted by the natives. He is the county chairman for the Queen Anne's Democratic Party, and owner of Fox's department store. In it Goldstein sells everything from construction boots and long underwear to lingerie and Afro rakes. "These people don't want progress," Goldstein says confidently. "They're basically farm people and watermen. The city people move over here and right away start complaining about the lack of sidewalks, streetlights, garbage pickup, and such. The farmers who've never had any of these services have no sympathy, because they know they'll have to foot the tax bills." Goldstein rang up a sale for a black woman who looked like life had not been too easy on her. "Life is pleasant and relaxed over here," he said. "There's none of that city pressure. They're happy with their lot in life. They don't have to work too hard."

Goldstein says the natives will treat you right if you deal with them openly and honestly. "If they like you, they'll do anything for you." But people here are very close, everyone is related, he adds, and if you bad-mouth someone, it gets around everywhere and you are finished. "They'll freeze you out."

———— ᙯ ————

A few years ago an obituary column in the Centreville paper read something like "Ninety-seven-year-old native of Baltimore dies here." The woman had moved to Centreville when she was about three months old.

———— ᙯ ————

Like every other Queen Anne's County resident, Sam and Lillian Whitby had received the Rouse brochure about Wye Island. Sam figured that the Rouse development was inevitable, but indigestible. "Well, you can't stop progress," he said, "they've got the money. But I would hate to see the island torn up like that."

Sam's reservations about the Rouse project, however, were not born of a desire to see the island preserved as it was. Sam did not share the feelings of the city people who had moved to the Shore and who were ecstatic to find an island so near the bay bridge and yet so "natural" and semiwild in appearance. Sam was disgusted at the way Wye Island had "all growed up." Most of the farmhouses of the two-dozen or so families who once lived there—Sam's Wye Island—had been burned away, bulldozed down, or lost in the impenetrable tangle of undergrowth and trees that have since been allowed to take over.

Sam tried to find them one afternoon. He drove past the head of Granary Creek, where he and Lillian had moved when they were first married. There was no farmhouse. He pointed out the window at the wild undergrowth and said, "That's the Robinson place." Farther down the island, the little house where his father had died: gone, overrun by honeysuckle and thorn bushes. Howard Melvin's former farm was buried in the same manner. Sam stood on his tiptoes, frustrated, trying to see anything identifiable from the past. "I can't see in there, but it strikes me that the house sat over there in the trees. There was a house here. Indeed there was. There was corncribs, and a horse barn. But they burned so many down to save taxes. Boy, how anything can grow up so thick!"

At the center of the island Sam drove down a barely passable lane into what was once the Bigwood Cove farm that he and Lillian had run for the Whaleys. Sam's schoolteacher had once boarded here. This would be Rouse's village. The yard was a sea of honeysuckle and dead trees. "That used to be a beautiful orchard over there," he said, pointing to a dense web of tangled undergrowth.

Sam stepped back and shook his head with dismay at the verdant jungle that he once knew as a neatly kept farmyard and at the wild hedgerows that towered over the road.

"That's one thing a farmer didn't do in those days," he said. "They kept their fence rows trimmed. Everything was as neat as could be."

On Wye Island near the bridge sits one of the few remaining farmhouses on the island. Like the Bryan house on Bordley Point, where Sam Whitby grew up, it, too, has long been abandoned. The house is used now only as a perch for buzzards. The yard around it can hardly be called a yard—it is overgrown so thick in weeds and brush. Gordon Shawn moved here in 1907, when he was about ten years old, and stayed on the island until the Depression. It was then called the Riverton farm, and it included the seventy-acre grove of widely spaced beech, sweetgum, willow, and white oak just behind it—Shawn woods—the largest woods on the island. Gordon Shawn's father raised wheat and corn, Guernseys, Holsteins, and a few hogs on the Riverton farm. The Shawns now live on a large farm called Bloomingdale, which is identified along the main highway by a state historical marker. The south wing of Bloomingdale was built in the late 1600s. The "new part," as the Shawns refer to the main section of the thirteen-foot-ceilinged house, was added in 1792.

Gordon Shawn's memories of the trim farms of early Wye Island do not fit well with his impressions of the island today. Of the Rouse plan for Wye Island, he says simply this: "Let 'em develop it. It'll be a good investment. I think something ought to be done besides what's being done down there now."

On a late afternoon Ray Warner sat in an old metal chair beneath the enormous beech tree that dwarfs his yard and small weathered house near Bryantown and talked about the days, just after World War I, when he managed the Whaley farm at Bigwood Cove on Wye Island. Ray Warner remembers that a pocketbook left on a fence post would remain until the owner finally found it. Ray Warner was in his thirties then—he's eighty-six now—and Ralph Whaley's father paid him seventeen dollars a month and provided a winter lay-in. He cared for nine cows and some sheep, with the help of his Chesapeake Bay retriever Beaver, who would drive the cows into the "pound" (Eastern Shore language for barnyard) at milking time. The dog never went in with the sheep, particularly when they were lambing, because he knew he would scare them. "Had more sense than any dog I knew," Warner said. The Chesapeake Bay retriever, a native breed of the Shore and a common sight in many yards there, is a powerful dog, capable of breaking through ice to swim to a downed duck. Warner used to raise Chesapeakes and compete in field trials. "Met some nice people at dog trials," he said. Warner has only one Chesapeake now. He named him Warner Sarge. "I used to give 'em names like Sarge O'Wye. I had a dog named Steve O'Wye. He was some dog. Bring the ducks right in the blind. I shot with one fellow who could never figure out how the dog could go right to the duck after he was shot, 'cause the dog couldn't see the water from the blind. I told him to watch the dog next time he was shooting. That dog would be watching where you swing, so he knew where to head when he came out of the blind."

For a moment Ray Warner stared at the road that leads down to Bennett Point. Warner Sarge began barking furiously from behind the house. Ray Warner got up slowly and walked around to the kennel. "Is he a good retriever?" I asked stupidly. "Uh-huh," said Ray Warner. From a small shed he produced a strange looking pistol that had holes in the sides of the barrel. He broke it open, inserted a single bulletless cartridge, and snapped it shut. He slid a white spongy sleeve, about the size and shape of a boat bumper, over the barrel, and locked it in place. Warner opened the latch on the kennel and the large chocolate-colored dog roared out, barking and leaping around Warner. "Sit," said Warner, and Warner Sarge sat, his yellow eyes fixed on the pistol. Warner stepped to one side and fired. The white retrieving dummy shot like a rocket far over Warner's large vegetable garden and almost to Arthur Bryan's woods, about 150 yards away. The dog was quivering like a bowstring. "Fetch," said Ray Warner. The ground shook as the dog thundered off to the trees, snatched up the dummy and returned with it held high.

Ray Warner still manages to go out for crabs occasionally in the summer. More rarely, in the winter, he hand-tongs for oysters with nippers. When Warner was young, he sold oysters at a quarter for three bushels. "Now they get five dollars a bushel. A lot of differences in the times," he said, putting Sarge back in his kennel. "I used to buy gasoline for seven cents a gallon. I been trying to buy a wood stove, but you can't buy 'em anymore." Like Sam Whitby, Warner built his own house. I asked what he thought of the Rouse Company's plans for Wye Island. "I think it's a shame to put a whole lot of people there. The Wye is one of the nicest rivers on the Chesapeake Bay." He turned and watched a car pass down the road. "Now if they did it at five-acre lots, that wouldn't be so bad," Ray Warner said.

————————⌀⌀————————

Hardy, and then Rouse, had hoped to buy all of Wye Island, but Arthur Bryan wouldn't budge off his 150 acres at Bordley Point. "If it hadn't been for me," he roared, "they'd have sold the whole island!"

Arthur Bryan was standing near the house he built in 1940, looking down the Wye River to the profile of Wye Island in the distance. An unseasonably cold March wind was blowing off the river below the bluff on which he stood, rattling the tree branches about him. He pulled up the fur collar of his greatcoat, and shoved his hands so deep in the pockets that the sides of the coat stood out from his thin frame like flapping shutters. Thatches of white hair protruded from under the flaps of the cap he had pulled down around his neck. Eighty-two years old, a confirmed bachelor, Arthur Bryan planted his legs wide apart, and his defiant eyes swept the high shoreline of his farm—the visage of a somewhat tattered lord of the manor surveying his domain. His reedy voice cut through the wind like a claxon.

"What you own—if it isn't land—isn't worth anything when you die! I own all sorts of bonds, AT&T, Treasury, and so on, but what are they worth? They're not worth *anything*! They're just a *promise* to pay. And *who's* making the promise? A bunch of *Democrats*!" he rasped. "*Money* isn't worth anything! The only thing worth anything is *land*!"

Arthur Bryan has never lived on Wye Island. His mother had so hated its isolation that shortly following her illness, brought on by the birth of Arthur's older brother Edgar, during the blizzard of 1888, the Bryans pulled out of Wye Island and moved to the farm at the head of Wye River where Edgar now lives. They left the Bryan place on Wye Island to be managed by various tenant farmers, like the Whitbys. Arthur Bryan eventually inherited it. Arthur Bryan intended to hang onto every acre of farmland he owned. If that meant slightly lousing up James Rouse's plans for the island, so much the better.

Upon first impression it might seem that here is a man who shares the late Aldo Leopold's philosophical opposition to treating land like a commodity, a farmer who views development as an act of spoliation, of sacrilege and desecration. That is not the case. To Arthur Bryan land means mainly one thing: wealth. He made considerable money in his younger days buying and selling it. He sold real estate during the day while he attended law school at night, and after his medical discharge from the army—having tried to cure double pneumonia, or "black flu" as he calls it, with a quart of Old Crow and a bottle of aspirin—Bryan joined a Baltimore law firm that dealt heavily in land transactions. He was soon making land appraisals for the city of Baltimore and acquiring and selling farms. Arthur Bryan was quite adept at convincing farmers to sell out. One day in 1920 he drove out to a farm on the edge of Baltimore. He found the farmer in a cornfield cutting the stalks by hand with a long knife. "I don't know how much longer I can continue doing this, but it's the only thing I know how to do," the farmer said. Arthur Bryan reminded the man what a burdensome life he was leading, and that if he would sell the farm he would have enough money to forget forever about droughts, early frosts, rapacious insects, and failure. The farmer wavered, but in time he folded and sold his farm to Bryan for three hundred dollars an acre, a very healthy sum in those days. Bryan's firm tore down the barns and sheds, turned the cornfields into a golf course, and later sold off the road frontage at a substantial profit.

The wind was growing colder. As he walked in the back door of his house, Arthur Bryan shouted, "I wouldn't sell to Rouse if he gave me five million dollars!"

Of Bryan's stubbornness over Wye Island, Frank Hardy once remarked, "He plans to take that with him."

———— ✧ ————

"We don't need a town down on Wye Island." It was Edgar Bryan talking. He sat on the screened porch of his farmhouse just two miles up the Wye

River from his brother's. In Washington, DC, an hour and a half drive to the west, it was the hottest May 17 since the nineteenth century. For such a suffocating day outside, the Bryan's porch was surprisingly comfortable. Mrs. Bryan brought in tumblers of iced tea on a tray with a plate of cookies and passed them around.

"It's just time to develop that island," she said as she set the tray on a table.

"I'm against it," her husband replied. "You're the only one who's for it." He paused. "No, I guess there is the fellow who owns that hardware store in Centreville. He's for it, too," said Edgar Bryan, smiling at his wife. Edgar Bryan is eighty-six years old. (When Edgar was about three months old, his parents had him christened on an undertaker's cooling board. His father, like Sam Whitby's father, had lived on Wye Island before there was even a drawbridge or rope ferry. The Pacas had a private bridge at the east end of the island, but they wouldn't let anyone else use it, so Bryan's father would throw a saddle in a skiff and row across the narrows with his sorrel horse swimming behind.)

Like most farmers, Edgar Bryan is proud to be associated with good farmland. "Wye Island had the reputation of having the best wheat farms in the state," he said. He shook his head. "It's a shame they tore down that old schoolhouse. No one cared."

Mrs. Bryan stood up. "Rouse is the one person who could do it right," she said. "He's a gentleman, he really is. He brought a plan in advance and said what he is going to do. People have got to live somewhere."

Edgar Bryan was reclining in a lounge chair. He turned his head and said sharply, "Let 'em go down to West Virginia and buy some of *that* land!"

Mrs. Bryan turned away from her husband. "The opposition to Rouse boils down to increased population, loss of game and shellfish," she said, and then she repeated what Frank Hardy had been saying publicly ever since the Rouse option was announced a year ago. "But it's going to be developed." She passed the plate of cookies around and then sat down again. She and her husband spoke without rancor, but the subject of the Rouse plan warmed them up more than did the weather.

Edgar Bryan sat up. "You take that entire stretch down to Bennett's Point. All those houses have septic tanks and most of them are fifty to sixty feet away from the river. Where do you think that sewage goes?" Edgar Bryan

73

asked, his voice rising. "Right into Wye River!" Mrs. Bryan reminded him that Rouse was planning to put in a waste treatment system for the village, and Edgar agreed that that would control sewage better than septic tanks on five-acre lots. "But it's boats that's the problem," he said. "I know what little bit of traffic there used to be. Now you go down to Wye River and the grandchildren are running outboards all over. And my grandchildren go down and raise just as much sam as any of them."

Edgar Bryan lay back and looked out through the screens at his lawn. He said quietly, "Everyone's running away from people, but there's no place to run now."

Mrs. Bryan looked at her husband and said in a soft voice, "It can't stay as it is. Mr. Rouse is the best developer."

"He's still a developer, Mom," said Edgar.

Mrs. Bryan persisted: "I'm interested in Wye Island because it's historical and beautiful, and I'm interested in it getting in the hands of people who will be sensitive about how it's used."

"Mr. Rouse is a fine man," said Edgar, "but he's a *developer*. And, buddy, when he gets in, it will be a lot different in five years than what it is now. Wye Island belongs in agriculture. I was hoping Mr. Houghton would buy it up and give it to Wye Institute. The federal government could pick up the tab on it and make it into a game preserve."

Edgar Bryan stood up abruptly and walked over to the screen, and the expression on his face mellowed. "Look, there goes a bobwhite walking across the grass. I love to see them walk across the grass."

We walked back through the comfortable old house and out onto the back porch. Edgar Bryan looked across his immense cornfield toward the row of trees along the head of the Wye River. Beyond was the old Mainbrace Farm which Frank Hardy, who is subdividing and selling lots there, calls Hickory Ridge Estates. The farm that Frank Hardy lives on is farther down the Wye. Between there and Edgar Bryan's farm is another Hardy subdivision called Governor Grason Manor. In mild exasperation Edgar said, "Frank Hardy has me surrounded." Then his voice firmed and his eyes sharpened, and Edgar Bryan, who has seen his native Eastern Shore change from an era when a good mule determined a day's travel, and who has seen the flow of city people spread across Kent Island, down the Wye River to Bennett Point, and along almost every stretch of waterfront he knows, summed up his feelings about how the Shore is changing and about Rouse's plan for Wye Island this way: "The most obnoxious thing on earth is a surplus of people." He was silent for a moment, thinking about the people from outside the Shore who were changing the type of life he had known so long. "You know one thing, most of the rich are very objectionable. Most of 'em are."

Edgar Bryan shoved his hands in his pockets. "I like *farmers*. I like the type of people they are."

CHAPTER 4

THE RICH

"The Eastern Shore used to be right nice," a native of the Shore once told me, "just farmers and watermen. Then they built this goddamn bay bridge, and now we're invaded by a bunch of fancy investment bankers. Rouse is just going to bring in more of them. Ordinary people are gettin' squeezed out."

The statement is true, but in implying that the Shore has only recently begun to attract the wealthy, it leaves the wrong impression. The rich have been there since the first European settlement on the Chesapeake Bay, when William Claiborne established a trading post on Kent Island in 1631. They sailed over from England to claim large amounts of land for tobacco plantations, and in the holds of their ships they brought indentured servants and slaves.

This division of wealth and class resulted from the system of land settlement established over the colony by the second Lord Baltimore, Cecilius Calvert. It was his father (and his heirs) to whom King Charles had granted proprietary interest over Maryland, but the first Lord Baltimore died before seeing his province, leaving it to Cecilius to establish the colony. Cecilius and his successors, the other Lords Baltimore, held many of the perquisites of a feudal landlord over Maryland, in return for which the British Crown was to receive all the colony's gold, silver, agricultural products, and two arrows annually. In his capacity as "Proprietor" of Maryland, Lord Baltimore deeded land to the planters subject to an annual payment—"quit rent"—in lieu of their feudal services. The sum, perhaps ten pounds of "good wheat" for each fifty acres, was rarely burdensome. To attract settlers (and their quit rents) Lord Baltimore began by offering every freeman a hundred acres, and a like amount for his wife and each child over sixteen years of age. By bringing over five additional men, a planter got two thousand acres. Anyone having a casual rum at a dockside tavern in England ran the risk of being sapped and dumped in the hold of a Chesapeake-bound ship. More than a few tobacco planters added to their acreage on the Eastern Shore in this manner. Gradually the price for this almost-free land went up, until after 1683 it could only be got by purchase (initially at a cost of two hundred pounds of tobacco for one hundred acres).

During the colony's first eight years, Lord Baltimore designated each thousand-acre holding as a *manor*, on which the lord of the manor would run his own court system. Lord Baltimore, however, never fully established a quasi-feudal system in Maryland. The manor judiciary—thirty-seven manors in all—was abolished by an elective county court system in 1642. Some estates on the Eastern Shore still bear their original manor names, and many more farms cling to the names they were given when first patented (deeded) by Lord Baltimore to the planter.

Thus was Wye Island first settled by the English planters in the mid-seventeenth century. Of the five patented parcels into which the island was initially divided, the name of only one—Drum Point—has survived the years since Thomas Bradnox was granted the first parcel in 1658. What Bradnox did with his three hundred acres, which he named Cedar Bradnox, is not known, although like every other bay settler he undoubtedly cleared it of its trees and planted tobacco. By 1662 one John Felton had acquired the four-hundred-acre Drum Point from Lord Baltimore, and William Stevens was patented a thousand acres of the island, which he called Stevens Choice. In 1657 a group of London tobacco merchants struck this bargain with Thomas Cary: take twenty servants, sail to the colony of Maryland in America, find a thousand acres of good soil, clear it, establish a tobacco plantation there for our trade, and the land will be yours. Cary did so, and in 1664 the Purchase, one thousand acres of Wye Island, was patented to him. Two years later the island's last remaining virgin ground was acquired by William Price, the three-hundred-acre Price's Hill. The difference between the size of the island then, 3,000 acres, and its size now, about 2,800 acres, is probably due to errors in wilderness surveying and erosion over the last three centuries.

As Price was picking off the last tract of the island, the growing influence of the immensely wealthy Lloyd family across the river at Wye House began to be felt. In 1668 Philemon Lloyd snatched up Cedar Bradnox on "the Greate Island in Wye River" and later acquired all but the Purchase.

Not until the 1700s would the island pass to the Chew family, and then to the grandest of its occupants, Judge John Beale Bordley and William Paca. Throughout the succession of its owners Wye Island would variously be known as Wye Island, Lloyd's Insula, Chew's Island, Bordley's Island, and Paca's Island. But the name Wye Island would stick.

A style of life evolved from the plantation days which has never fully disappeared from the Eastern Shore: "The Land of Pleasant Living" it has come to be popularly called. It grew out of a society where those of the better plantations lived in opulent style, and for whom the best things in life were often gained with a minimum of effort. The work on the farms—and hard, sweaty, backbreaking work it was—was done by the slaves and field hands. Among the landed aristocracy there were people like John Bordley and William Paca whose achievements were obviously notable, but there were also those who achieved only heights of indulgence. To an outsider, it is a curious phenomenon why the descendants of so many not-so-notable rich ancestors busy themselves constructing elaborate genealogy trees from such sparse bushes, but they do. And from their efforts have come many romanticized histories of the Eastern Shore. One such outsider was a writer for *Harper's New Monthly Magazine*, who took the train down the Eastern Shore in the summer of 1871. Here were his observations:

Until very recently, immigration has been practically discouraged, not alone through the hostile sentiment of the old proprietors, but also their reluctance to part with any of their large (and generally encumbered) estates. It has been held to be more "aristocratic" to possess a thousand heavily mortgaged than a hundred free acres. Large estates belong to "blood," which is still a word of great potency in this world. I don't know how many times during the day the birthplace of somebody's grandfather was pointed out to me. Utterly unknown names and genealogies were explained with a patience and an enthusiasm which presupposed the profoundest interest on my part.

The perspectives of Arthur Bryan and Sam Whitby about their ancestral pedigrees could not differ more. Bryan traces his ancestry to a great-great-grandfather who came to the Eastern Shore from England, as did most of the first settlers. Bryan is proud of his English heritage and makes little effort to disguise his opinions of other nationalities. "The English," he says, "are the only fine blood that came into this country—not like them goddamn Germans who have bristles on their backs! The leaders in America didn't come out of Italy and Germany and those other little countries in Europe. It was *England*! They had the experience. They knew how to make good furniture, leather goods, and silverware. My parents were high principled with high qualities of generations behind them. It was wonderful. My mother painted pictures in oil. My father was nothing but a fine, high-class country gentleman. He kept his servants so well in luxury, well, you never had to lock up anything around here. He never had to do any manual labor. He farmed on a grand scale. He kept big herds of cows, sheep, and pigs and rotated his crops, using the natural fertilizer. Now, if you need anything done, you got to do it yourself."

"How did your father make all his money?"

"He *inherited* it."

When asked whether he knows anything about the origin of the first Whitbys who landed on the Eastern Shore, Sam Whitby says, "Don't have the slightest idea. My father always told me they were Irish people. I've watched a lot of television, and I've never seen a Whitby on television."

———— ⌁ ————

The first planters settled on the waterfront lands, since just about everything moved by ship or boat then. The earlier roads were nothing more than packed paths from the planter's drying shed to his wharf. They were made by huge hogsheads of tobacco which the slaves rolled directly to the ships—hence "rolling roads." Gradually, as the interior was cleared for tobacco, a few primitive horse and carriage trails were cut through the woods. Occasional towns began to grow up where the streams could be crossed—at the head of tide—and the roads through the back country were used more frequently. As these roads improved, the better soils were discovered inland, and the plantations spread all over the Eastern Shore. But well into the early twentieth century, commerce moved mostly by water, and owning land on the water was still more an economic necessity than an esthetic choice—though it was certainly cooler there.

The abolition of slavery broke up most of the larger plantations, whose owners were unable to handle their big estates when the slaves walked away. Like most of the South—which the Shore was alike in most ways—the Eastern Shore slid into economic stagnation, and farms were auctioned for a song.

In the 1920s and 1930s the Shore was rediscovered, this time by families of wealth and influence from the industrial eastern cities. By then the road system allowed farmers to truck their grain and produce to the waterside towns, rather than having to depend on boats sent from every private wharf out to the waiting ships, and they were willing to give up the waterfront for better inland soils. But the Great Depression turned a trend into a crisis. The Centreville bank was closed, and people were scared. The banks and the feed and seed companies foreclosed mortgages, and farm after farm either was auctioned off or sold by its floundering owner under distress. When the wealthy arrived to buy waterfront estates, they found thankful sellers. Arthur Kudner, who already owned a huge cattle ranch in New Mexico, bought enough farms along Eastern Bay to have almost four miles of waterfront—to which he commuted from his New York City advertising agency in his sky-blue Sikorsky amphibian. Out of Saint Louis came the president of the Carter Carburetor Company, Parke Sedley, to buy up the Shipping Creek Farm on Kent Island. The president of Pittsburgh Steel, Homer Williams, bought Marengo, just above Long Point on the Miles River; and the Wye Plantation was acquired by the head of Steuben Glass, Arthur Houghton, Jr.

And so it went throughout the 1920s and 1930s. Gradually the natives gave up the waterfront to the rich. Unlike the early planters, the city millionaires bought their waterfront estates for purely esthetic reasons. Here on the Eastern Shore they lived in genteel luxury, looking out across sloping lawns to their yachts on the water. In the winter when the ducks drilled across stormy skies, the gentry, in their shore blinds, were within walking distance of their eggs Benedict. The Land of Pleasant Living.

The effect of the rich on the upper Shore is felt today. They made the shorelines more beautiful. Crumbling old manor homes, gutted by fires or abandoned, were restored; the weeds were ripped out and landscaping put in. They pumped money into the local economy and gave jobs to carpenters, like Sam Whitby. The natives could point to these handsome places, and with a degree of pride and awe, mixed, for sure, with envy, say to visitors, "See that big place on the end of that point? Uh-huh, the one where that sloop's tied up. You know who owns that? The head of the biggest steel outfit in Pittsburgh. Indeed, it is. Whoooeee, y'oughta see the fancy Packard limousine they ride around in!" In 1937 the *Queen Anne's Record–Observer* proudly reprinted this editorial from another Shore paper:

The Delmarva Peninsula is studded with the handsome homes and broad acres of well-to-do families. . . the native population is not being supplanted. Far from it. It is merely being augmented by a wholesome and welcome influx of most delightful persons from here, there, and everywhere. . . . Just when and where this influx will end nobody can tell. It will be good if it never ceases. Only benefit can flow by winning as residents and homeowners good men, women, and children who love in advance the countryside watersides on which they settle.

Not all the rich, however, were welcomed by the natives, nor all the natives by the rich. In purchasing their vast estates, the wealthy were buying privacy and, in a sense, walling off themselves. They do it today on the Shore, as do others not so rich who retire or own second homes there. As do most people in this country who buy land. For reasons good or ill, owning land is the most effective way in which people keep their distance from others. Land is the ultimate means of exclusion. And to Wye Island in the 1920s, long before Jim Rouse tried to breach a broader wall of resistance, there came a man and a woman who used their dazzling wealth to serve an almost macabre exclusivity. They gradually bought up most of Wye Island (the little Bryan place and Wye Hall Farm excepted) and evicted the tenant farmers, until by the mid-1930s only Sam and Lillian Whitby were left. Then they forced out the Whitbys. Their land acquisition would indirectly set the stage for the Rouse option years later.

For more than forty years the brooding presence of Jacqueline and Glenn L. Stewart hung over Wye Island and their nearby castle. To many natives of the upper Shore, it remains today.

———— ✺ ————

"Mrs. Stewart? Sure, I remember her. She was one tough woman! Glenn Stewart was big. Wore a black patch over one eye. Had a big scar on his face. . . told me he got that in a duel in Heidelberg, or somewhere over there. He was an important diplomat with the government. Jesus, did they ever have the money. Never mixed much with the local people here, though. Understand they entertained a lot of dukes and princes down at their castle on Miles River neck. But folks here never saw much of him. He spent most of his time on his yacht—a real romantic adventurer. I think he was afraid someone was after him. They didn't want no one down on Wye Island." Thus go the typical recollections of those who came in rare contact with the Stewarts.

Jacqueline Archer Stewart was born in Ireland. She attended private schools in Paris and moved to Manhattan as a young woman. Jacqueline was apparently wealthy in her own right when she met and married Glenn Stewart. Stocky in appearance, she wore her hair in the fashionable bob of the twenties era, wrapped herself in furs, and bejeweled her ample bosom. She loved money, dogs, and horses, in roughly that order. To her marriage with Stewart, Jacqueline brought her money and the family coat of arms, a centaur firing an arrow over its rump.

Glenn L. Stewart was born in Pittsburgh in 1884. From his parents he inherited a fortune. He dabbled at Yale and Harvard, caromed about the world for a while, and went into the Foreign Service in 1914. He was a large man—six-feet-four, weighing 250 pounds—who affected a long gold cigarette holder and a pointed moustache like that of Agatha Christie's Hercule Poirot. He had a patch over one eye and a scar across his cheek, but neither was received in a duel. They were the result of a bomb—Stewart's own bomb, in fact. While at Yale, he was maddened to discover that some girls whom he had invited to a party had decided to attend another party out of town. Stewart constructed a bomb. He intended to blow up the tracks before the girls' train departed, thereby hoping to divert them to his affair. Fortunately, for both the girls and the railroad, but unfortunately for Stewart, the bomb exploded prematurely, blinding him in one eye and scarring his face.

Glenn Stewart's diplomatic career was anything but that. He held minor positions as a fourth-class secretary to the U.S. legations in Havana and Guatemala and the embassy in Vienna. Despite his low rank, Stewart freewheeled as an ambassador-at-large. When he arrived on station, his first act was to go off on ship cruises for a month or two, or three. For an entire year an exasperated State Department did not know his whereabouts. It is difficult to understand how Stewart stayed in the foreign service as long as he did, for he was not only blinking in and out of view like the Cheshire Cat (mostly out), but it is not clear how he earned his pay. A report he wrote on Guatemala was passed along by his supervisor to the head of Latin American affairs at State with this covering note: "[This] contains no information, whatever, of value to the department other than that which it has known for some time, with the possible exception of one or two of the tabulated enclosures, which bear the earmarks of being parts of consular reports. . . . It is without exception the most careless and almost illiterate document I have ever seen."

About his own financial affairs Stewart was similarly loose, and his creditors fruitlessly chased him by mail from legation to legation. For his steamship fare home from Austria, Stewart put the touch on the American minister to Switzerland. The minister soon joined the creditors' posse. But Glenn Stewart was cool under fire. At the very time the State Department was being hounded by his creditors, Stewart wrote the secretary directly, demanding full reimbursement for his travel expenses home, given the exigencies of World War I, rather than the statutory five cents a mile. In 1920, having shown a patience that is nothing less than remarkable, the State Department sacked him.

On their honeymoon around the world, the Stewarts stopped over in Granada, Spain, to soak up the exquisite Moorish architecture of the Alhambra palace. They liked the Alhambra (and royalty in any form) so much, in fact, that they decided to build their own castle and live in it, which was precisely what they did.

In 1922 the Stewarts moved down to the Eastern Shore and bought a point of land on the Miles River across from Saint Michaels. The Stewarts renamed the land Cape Centaur, and on it began to construct their not-so-little replica of the Alhambra. They imported Tunisian tiles for the castle interiors and large roofing tiles from Cuba. For authenticity, the plasterers had to hand-rub the stucco into the walls and vaulted ceilings, much the way the slaves constructed the Alhambra for Mohammed ibn-al-Ahmar and his successors in the thirteenth and fourteenth centuries. The floors were fastened with wooden pegs. Glenn Stewart was fond of telling visitors that the floors had been walked upon by almost every famous person in the world, an assertion of some truth since Stewart had got the flooring out of the popular old Shoreham Hotel in Washington, DC. Jacqueline's large bathroom was papered with eighteen-carat gold leaf. Affixed to her bedroom walls was a continuous canvas mural, painted by Victor White and once displayed in the National Gallery of Art in Washington, depicting the bloody adventures of Cortez. In the Cortez room Jacqueline was somehow lulled to sleep by gazing at scenes of horses and soldiers falling over cliffs and Aztecs holding the dripping organs of human sacrifices high in the air.

Centaur Castle never quite captured the delicate arabesque of the Alhambra. Instead, there was about the place, and the Stewarts, a darkness that suggested something fearful. This may be due in part to their taste in interiors. However, it is more probably because what the Stewarts built at Cape Centaur was less a romantic castle than it was a fortress, a place to hide. Glenn Stewart was convinced that someone was trying to kill him.

While stationed in Guatemala, the Stewart's quarters had been burglarized eight times. One of their dogs was killed, and two were stolen along with their chickens, tools, and most of the machinery out of the pump house. One midnight, awakened by an alarm, Stewart, clutching his pistol, stumbled into the pump house to find four Guatemalans trying to remove the pump itself. Two of the men fled, but the others came at him with a knife and an iron bar. Stewart killed both. Stewart was advised that blood revenge was a common practice in Central America, and that unless he caught the next boat out he might be ambushed. Glenn and Jacqueline Stewart were on the next steamship.

The Stewarts were obsessed with making Centaur Castle an impenetrable stronghold—so much so that they allowed no workman to do more than a small segment of the construction. One of Lillian Whitby's cousins was employed as a painter, but he was permitted in only one room. The doors were of half-inch steel plate sandwiched between thick slabs of solid oak, and secured by large Fox police locks. One entire wall of Glenn Stewart's dressing room consisted of drop-hinged cabinets, each drawer with a separate lock. Most of the castle windows were narrow slits about four feet high—just enough space through which to poke a rifle. The spacious arched windows were protected by interior shutters of quarter inch steel, which could be swung into place and locked; a narrow steel panel in each shutter could be snapped open and fired through. The walls were almost three feet thick. On the roof Stewart installed an air raid siren, which could be activated from the tower. It once went off accidentally and spooked the neighbors. The county made Stewart disconnect it.

But the Stewarts were still fearful, so to further seal themselves off from attack, they added a three-story tower to their bastion, which was connected to the main hall by an arched passageway. On the top floor of the tower, among a museum full of Chinese antiques, slept Jacqueline (when she was not inclined to slumber with the battling Aztecs and Spaniards in the Cortez room). Glenn Stewart dozed fitfully on the second floor. On the main floor of the tower was Adolph Pretzler, their Austrian bodyguard, Glenn's personal secretary, and general manager of the Stewart holdings. Both Stewart and Pretzler slid loaded Colt .45 revolvers under their pillows. Each evening, Glenn Stewart opened a compartment in the wall of the circular staircase outside his bedroom, reached in, and switched on a concealed motor. The tower hummed and clanked, and, from his cot below, Pretzler watched a huge portcullis of bolted, gridlike four-by-fours descend through a shaft in the archway until the pointed ends rested on the floor. In this medieval manner the Stewarts retired for the evening, nagged by the thought that a team with grappling hooks and rope ladders might yet breach their last circle of defense. If during the night Glenn Stewart heard a noise, any noise, he would arouse Pretzler, lift the portcullis, and send him into the night. While Pretzler stumbled about in the dark with his flashlight and pistol, Stewart, clutching his trusty Colt, climbed to the ramparts leading to his turreted study over the roof. He would call to Pretzler far below: "See anything?" Pretzler never saw anything—no Guatemalans in muffled canoes, no commando teams with grappling hooks, nothing. Pretzler would grumble an all-clear and head for his cot. But he had to wait for Stewart to raise the portcullis. Pretzler's job had certain drawbacks, but he was patient. And his patience did have its rewards.

Around Cape Centaur the Stewarts threw up their outer defense perimeter. At the road entrance they built a pair of heavy, timbered gates, duplicates in strength and appearance to the portcullis. The gates were padlocked by a heavy iron bar when closed—which was all the time. An armed sentry guarded the gate from a brick turret. No one was allowed in or out of Cape Centaur without a written invitation and exit card, dated and time stamped. The guard once prevented a tractor salesman from leaving because the time had elapsed on his exit card, and he was sent back to the castle to have his card properly stamped. A high, wire fence encircled the estate. Within it a wide strip was cleared of all foliage and trees. Large Irish wolfhounds and stern men on horseback, cradling shotguns in their laps, constantly patrolled the fence line. It was widely believed among the neighboring shoremen that if you touched the fence, it would fry you like a potato chip.

The closest that the Stewarts ever came to an attack on Cape Centaur occurred when two dark figures quietly swam around the fence and crawled onto the beach. A police whistle shrieked from the tower, and two wolfhounds—a species bred, since 300 B.C., to dismember timber wolves—pounded across the cornfield. One of the saboteurs sprinted to the water and swam safely away. The other leaped for the fence, vaulted it, and fled through the trees like a panicked deer. The two were twelve-year-old Boy Scouts, intrigued about the mysterious Stewarts and looking for a way to spice up their camping trip. The one who had demonstrated that the fence was not electrified was the champion of last tag: Jimmy Rouse.

Jacqueline Stewart was never without her dogs. In her arms she carried two poodles whose coats she dyed each year to match her latest Cadillac's interior. At one time her kennels on Cape Centaur held thirty-six Irish wolfhounds (she gave two to Rudolph Valentino). Jacqueline's kennelmaster lived in a specially built home in the middle of the kennels; he fed the dogs about 175 pounds of beef a day, along with immense amounts of cornmeal, homemade bread, and rice. Jacqueline wrote (in the *American Kennel Gazette*, June 1925) that the wolfhound "is a rich man's dog. . . . A poor man should not attempt his care, any more than he would try and keep a Rolls Royce on a Ford car income." Her favorite was a 190-pound giant called Bally Shannon, the largest Irish wolfhound bred in modern times. Jacqueline bought him for $1,250 and immediately hired him out, charging $500 for stud service. It was generally accepted among dog breeders then that good old Bally Shannon was hauling down a higher stud fee than any other dog in the country. He made $10,000 for Jacqueline before he rolled over and died seven years later. Jacqueline Stewart had him stuffed and shipped to the American Museum of Natural History in New York. He was displayed there for years, along with five other wolfhounds who gave up the ghost at Cape Centaur. Today Bally Shannon is only pieces of skin kept in a large plastic bag in the museum's storage area.

After Jacqueline Stewart ran out of Irish wolfhounds, she bought (in China) some muscular, unfriendly chows. As a boy, Jim Rouse remembers seeing the Stewart's Stutz Bearcat parked outside the bank in Easton. He peered through the windshield. Fortunately for Rouse the windows were bulletproof, because what next exploded against them were the fangs and purple tongues of two roaring chows bent on pulling their small inquirer into an infinite number of pieces.

One of the chows went everywhere with the Stewarts—and the Stewarts went everywhere. They spent less time on the Shore than away from it. Not about to crate it in the baggage car when she traveled, Jacqueline dressed the chow in children's clothes, bundled it in a blanket, and had Glenn carry it like a snoozing child into the Pullman. For their silence, the porters made a small fortune in tips from the Stewarts.

Pretzler accompanied them on most of their trips, not just for protection but to carry emergency provisions as well. He carried these in an aluminum suitcase. Four holes were drilled in the bag to keep the contents from generating combustion and catching fire. The contents of the bag, the emergency provisions, consisted of cold cash, more than $500,000 on one trip. The Stewarts wanted to be able to dip into it should the castle be attacked in their absence, forcing them to flee with only their journey bags. But this was chicken feed compared with the amount they stashed about at Centaur Castle: at one time no less than $1.6 million in paper currency and silver and gold coins.

Among the fantasies that fluttered through Glenn Stewart's mind was the vision of one day owning the entirety of Kent Island. "I am the Duke of Kent," Stewart would say, as Pretzler chauffeured him about in the Duesenberg.

Stewart would refer to Pretzler as the Crown Prince or the Count of Leobenn. (That was the town near Vienna where Pretzler was born.) Stewart never fully realized his feudal, or futile, dreams, but he and Jacqueline did not do badly. In addition to Cape Centaur and their castle, they bought four more farms on Miles River neck and a 3,500-acre cattle ranch near Conifer, Colorado, converted an old Easton hotel into an office building, and gradually began buying up Wye Island.

Unlike Judge Bordley, the Stewarts knew little about farming, and, given their propensity to travel for months at a time, they had difficulty bringing off their various agricultural plans for Wye Island. They tried raising Percheron horses. That failed. They next shipped in three thousand sheep from Montana and hired a university professor to manage the considerable flock. They lost money on the sheep. As a last resort, Jacqueline Stewart trucked in Hereford cattle to Wye Island from Colorado and Kansas City (it appears Glenn had little interest in running anything), and she hired western cowboys as herdsmen. With the purchase of each farm on Wye Island, Jacqueline's cowboys would strip out the hedgerows, fence the fields, and run the cattle in. The cattle stayed.

Glenn Stewart enjoyed gunning, not ranching, though he sometimes wore a white linen suit and cowboy boots. Stewart looked upon Wye Island as his sporting hideaway, as well as another hiding hideaway. On Granary Creek, he had constructed a brick hunting lodge and paneled it in knotty pine. He called it the Duck House.

The Duck House was no ordinary shooting lodge. But then Glenn Stewart was no ordinary marksman: he had already bagged two people. Stewart considered the possibilities of a night raid on the Duck House. The thick oak doors and window shutters, like those in Centaur Castle, were bulletproofed with steel plate. A large cement basement was built under the Duck House, but it had no windows, no doors, and no stairs—offering no evidence to the outside world that there was anything under the Duck House except solid earth. Access to the basement was possible only through a false floor in front of the living room fireplace. In the same secretive manner that he had constructed the castle, Stewart divided the Duck House work crews, sending Pretzler out to get a large hydraulic lift from a filling station for raising and lowering the floor section (and the foot-thick cement slab that it rested on). As Hitler rolled over Czechoslovakia and Poland,

Stewart's fears of Guatemalan ambush grew into a terror of panzer invasions of the Eastern Shore. He prepared his basement for a long siege: he stocked it with a twelve-month supply of food and staples, a flour mill, a bed, ample clothing, and plenty of ammunition. He painted the roof of the Duck House and the walk around it in mottled camouflage to blend with the surrounding trees. Dive-bombing Stukas would have a hard time finding Glenn Stewart.

For all this effort Stewart never stayed in the Duck House, preferring, instead, the mobility of his shadowy yacht. Jacqueline and Glenn tended to go their separate ways, and during much of the Stewart reign on the Eastern Shore Glenn Stewart was off sailing in South American waters, trying to trace Columbus's voyages. The yacht *Centaur* was no small sailing vessel. It was an eighty-foot schooner. It was painted black. To captain the *Centaur*, Stewart hired Al Capone's skipper. All that it lacked was a skull and crossbones. Glenn kept Jacqueline fearful. As they walked around the shoreline, he would muse aloud, "You know, my dear, a body could be slipped into one of these coves and never be found."

To Jacqueline's relief, one day Glenn Stewart sailed away to Nassau and never returned. The Stewarts later divorced, and Jacqueline gained all the Eastern Shore property. (Glenn Stewart remarried and is thought to have died in the 1950s.) But long before Glenn Stewart at last drifted off, Jacqueline was in command of their vast estate. Her armed horsemen and fleet wolfhounds patrolled Wye Island constantly, and from the day Jacqueline first set foot on the island word passed quickly up Wye Neck that trespassers were unwelcome.

———— ♻ ————

Jacqueline's vise gradually tightened. Some local watermen, Sam Whitby included, had been mooring their workboats in a cove behind Drum Point, to be nearer the mouth of the Wye and Eastern Bay beyond. But when Mrs. Stewart—the local people knew her by no other name—bought the Drum Point farm, she closed off the farm lane that connected the cove to Wye Island's only public road and threw the watermen out. Enraged, the watermen hired lawyers and sued to open the lane, but the court had no choice—Mrs. Stewart owned the land—and the watermen were evicted. Then she shut down the Drum Point rope ferry, and even tried, unsuccessfully, to close the island's county road. By the mid-1930s Jacqueline had evicted most of the tenant farmers on the two-thirds of the island she had acquired. The schoolhouse on Dividing Creek stood empty. She bought it from the county and filled it with hay.

Sam and Lillian Whitby were now virtually alone on Wye Island, farming the Whaley place on Bigwood Cove. When Mrs. Stewart bought that farm, she told the Whitbys that they could stay there but no longer farm it. Sam and Lillian moved to the last place left, a small farm nearby. The Whitbys could feel Mrs. Stewart closing in. "I have money. I have power!" she warned Sam.

———————∽∼◦———————

Her presence was almost palpable, for Jacqueline Stewart splashed herself with a perfume that spread like a tainted fog. One afternoon Sam was driving a team of mules down the island into the teeth of a wind so fierce that it soon obscured the road in billowing dust. Through the dust Sam caught the scent of Mrs. Stewart's perfume. It grew so pungent that he was afraid the blinded mules were about to run her down. He halted the mules, and suddenly the wind shifted and blew the dust across the field. Jacqueline Stewart was standing in the middle of the road. Three hundred yards away. *"That's* how strong her perfume was!" Sam said to me one morning, as we drove up to the Duck House. Hardy leases the Duck House each year to Rockwell International, who brings customers and government officials down to shoot geese on Wye Island. When Sam and I arrived, some of Hardy's farmhands, who guide Rockwell's shooting parties, were carrying goose decoys down a stairway that Hardy had built into the basement. Sam had never been there. We descended into Glenn Stewart's never-used bunker and stood in a semicircle around a glistening steel cylinder that reached from the basement floor to the concealed floor section over our heads. I asked one of the men whether he remembered Mrs. Stewart. "Oooh, yea," he laughed. "Indeed I do! You could smell her a mile away!"

Sam swung around to face me, his account of Mrs. Stewart's powerful perfume confirmed. "See *there?*" he said.

———————∽∼◦———————

For the Whitbys, life on Wye Island had become a tense drama. Jacqueline Stewart was mercurial; she would be friendly one day, and on the next, confronting Sam in the field, would say to him, "We're going to *buy* that farm and *you're* going to work for *us.*" Sam told her he would not work for her, and he continued plowing. "I'll *get* you off!" Jacqueline warned. She bought the farm that Sam was renting and told him to get out. Sam told her that sort of thing was not done on the Eastern Shore, that unless the lease had been cancelled before January a tenant could stay on the farm until the end of the year. She threatened suit, and Sam hired a lawyer, but he felt all alone. "She was one very ornery woman. She didn't want no friends. All she wanted was her money. Her lawyer tried to collect half my wheat. But there was just me and her cowboys living down here at that time, and I was afraid I might get killed. That's when I left for good."

Jacqueline Archer Stewart died in 1964, leaving behind a tangled estate worth millions. In probate Adolph Pretzler produced a handful of papers signed by Jacqueline leaving virtually all of the estate to him. The court refused to accept the unwitnessed documents as proper wills, but Jacqueline's nieces and nephews finally agreed with Adolph to settle. Pretzler got 28 percent of the estate, including Cape Centaur, where, with his lovely German wife, he lives today.

The appraisers went to the castle and began compiling their lists: sable and mink coats, gold watches, gold penknives, a live-steam toy locomotive, Tiffany bowls, a framed mezzotint, hundreds upon hundreds of pieces of silver flatware, trunks full of jewelry, diamonds, emeralds, a '31 Duesenberg convertible, a '24 Packard straight eight, three more cars, a sleigh, a Steinway grand *player* piano, a crossbow, and on and on. A rumor came to the appraisers that there might be more. They asked Pretzler whether they had missed anything. Pretzler showed the appraisers into Glenn Stewart's dressing room, and opening a cabinet, he manipulated some concealed levers. Suddenly, a section of the floor yawned open to reveal a basement stairway. The men walked down the stairs. In the basement they found bushel baskets and grain sacks full of jewelry and coins, over $6,000 of silver dollars, fifty-cent pieces, quarters, dimes, and pennies—tipping money. But the find that took their breath away was a pile of ten- and twenty-dollar gold pieces—more than four hundred gold coins in all—some minted as far back as 1850. When all the cash found in the castle, including the basement bundle, was totaled, it came to $160,000—and that at gold prices far lower than today's.

The Stewart portion of Wye Island was sold at a sealed-bid auction in 1965 to Frank and Bill Hardy. The Stewart era was ended.

Yet the memories linger on at Cape Centaur. Only a few weeks after Rouse had placed his vision for Wye Island upon the library table, I drove out to Centaur Castle. Jacqueline's crest was still embedded in the brick gate posts, though a sentry no longer paced the ramparts of the turreted guardhouse. Deep within the fenced perimeter the pink castle loomed in old age. A high iron fence encircled it. Low, gray clouds put the castle in shadow, and a rain began to fall. Adolph Pretzler was home. As we talked in the "great hall," the sloughing coals from a distant fireplace gave a flickering light. The stucco was loosening in places along the vaulted ceilings. A corner of the canvas had begun to peel from the wall in the Cortez room. The portcullis hung in its shaft, but the motor to raise and lower it did not work. Of the Rouse plan, Adolph Pretzler said it would be awful to see all those homes on Wye Island.

Jim Rouse, or his associates, had met with over fifty local civic, watermen, and farming groups, but he worried as much about getting support from the rich landowners around Wye Island—Shep Krech, Tom Wyman, Eugene du Pont III, Arthur Houghton, Jack Kimberly—and the rest of the upper Shore gentry. Rouse wrote letters to people like Russell Train, the administrator of the Environmental Protection Agency, who owns a large farm below Saint Michaels, and he hauled his scale model down to the office of the then Secretary of the Interior Rogers C. B. Morton. Morton's waterfront estate, Presqu'ile, lies on the Wye East River, directly across from Wye Island. Before Morton bought Presqu'ile in the early 1950s, the land around it had been subdivided and sold off in five-acre lots. Morton knew that under the present estate zoning that was precisely what his view of Wye Island would turn into: here a dock, there a dock, everywhere a boat dock. He liked Rouse's concept of deep setbacks from the water and no private docks around the shore. But Rogers Morton thought Rouse was planning to put too many people on Wye Island.

Among the rich, Arthur Houghton, of Wye Plantation, was the closest to being an outright supporter of Rouse, but he did not think it proper to take a public position on the plan until Rouse formally applied to the county for the rezoning—nor until Houghton could review the architectural details for the village. Rouse found Houghton willing to debate the pros and cons of the plan; with most of the others, Rouse either encountered antagonism or no reaction at all.

It was a frustration with which Houghton was sympathetic.

"You'd think it was the end of the Eastern Shore to hear our friends in Talbot and Queen Anne's counties," he said. "We have spent over a year with our neighbors saying to them not to prejudge Rouse's plan until we see whether it is what we want or don't want. It's just that they don't have open minds." Arthur Houghton sat at the conference table in the visitors' center of Wye Plantation. He wore a soft yellow shirt, open at the neck, maroon slacks, and suede loafers. He wore no socks. A white handkerchief was folded in the breast pocket of his brown, double-breasted jacket.

Houghton folded his arms and leaned low over the table. "Here on the Shore," he said, "there are the 'old' families—and I put that in quotes—who still have some money left. And then there are the wealthy people who want to help put up the barriers to change, who have moved in seeking the good life with mint juleps on the veranda and black servants in white coats. I've noticed that the new money and the old families mix together. You know, 'They're *our* type of people,' they say. Private schools, travel, that sort of thing. You get that together and you have got a pretty powerful thing."

Seated next to Houghton was Jim Nelson, the executive director of Houghton's Wye Institute, and himself a native of Somerset County on the marshy lower Shore. In Somerset a fourth of the work force was unemployed. Nelson said, "You have people who have masterminded the biggest oil and insurance companies in the country, but they get down here and they shift to the opposite side and oppose everything they have stood for."

Arthur Houghton continued. "Change is going to take place," he said. "We set up this Wye Institute to be a few steps ahead, to assist—the word 'assist' is important—change, so that we can hold on to the best of the past, but, more important, get the best out of the future. I don't think any community can make a bigger mistake than to stop progress. You can only delay it for a while."

It was past noon, and Arthur Houghton invited me for lunch. I drove down a long dirt road between Sam Whitby's split-rail fences and circled the white pebbled drive in front of the Houghton mansion. To one side, in a small, manicured grove, was the grave of William Paca, signer of the Declaration of Independence. Across the drive stood the Houghton library, faithfully recreated to look like the original house of William Paca that Houghton pulled down when deterioration made restoration of it impossible. In the den Mrs. Houghton served sherry in Steuben crystal. Their houseguests included an Englishman who was availing himself of Houghton's unique library to study the poetry of Sir Walter Raleigh. From time to time Houghton would disappear into an adjoining room to read the wire service ticker that chattered quietly behind thick walls. Lunch included vegetables from the Houghton's large garden. Lunch was served by black servants in white coats.

———— ✺ ————

Bradford Smith, Jr., first saw the Eastern Shore from the railing of his yacht. In the summers the Smiths left Bryn Mawr and sailed the Chesapeake Bay. One anchorage they favored was Reed's Creek, near the mouth of the Chester River. In 1961 the Smiths bought Reed's Creek Farm. Two centuries ago Reed's Creek House had been a grand mansion; it was built before the American Revolution by a colonel in Washington's army. Fox hunts had begun on its lawn (the fox was dumped on the grass from a sack and allowed a head start on the hounds). Throughout the twentieth century, when the farm was a dairy, the mansion deteriorated. The dairyman and his family could not afford to keep it up, so they closed off all but three of the rooms and huddled around the fireplaces and, later, space heaters.

Using local carpenters, the Smiths invested a small fortune in restoring the mansion over a nine-year period, beginning by installing its first water heater. Under centuries of dirt and grime they discovered a painting on the paneling over one of the many fireplaces. Smith brought in restoration experts from du Pont's Winterthur. After patient cleaning they uncovered one of the earliest American nudes (1793): a scene from a fifteenth-century poem depicting Armida seducing Rinaldo in her garden. The artist's signature, W. Clarke, was visible only in infrared light.

When Brad Smith finally retired as chairman of the board (and founder) of the Insurance Company of North America, he and his wife gave up Bryn Mawr for good and, on New Year's Day 1970, moved into their restored Reed's Creek House. Recently, the mansion was placed on the National Register of Historic Places.

No longer busy managing a large corporation, Bradford Smith enjoys his new life puttering about the estate, doing things for himself. He likes rubbing elbows with carpenters and farmers. "I don't want to be misunderstood," he said one chill[y] afternoon by the fire, "but in Bryn Mawr you didn't *know* the undertaker." Although Brad Smith is a member of the Corinthian Yacht Club (Philadelphia) and the Merion Cricket Club (Haverford), he seems to have gotten as much pleasure from being accepted into a local rod and gun club. He speaks of the natives with warmth: "Oh, some of them speak a little roughly, and a few really murder the English language, but they are a fine, gentle people. They may not have any money, but they are certainly helpful." When some of his Philadelphia friends hear of Brad Smith associating with bricklayers and crabbers, they suggest that he must be slipping a little. Bradford Smith loves the rural atmosphere of the Eastern Shore. It delights him to see his grandsons walking barefoot up the lane, carrying a bass they have caught in the pond. He wants the Shore to stay exactly as it was when he first found it.

Brad Smith is convinced that growth and development of the Eastern Shore is not inevitable. Just to be sure, he formed a local conservation association to fight any major development—like that of Pioneer Point, the old Raskob estate that lies across Reed's Creek from Smith's mansion. As he watched a flock of geese circle and drop into the creek around his pier, Smith said, "One of the things that puzzles me is these people who make the assumption that Queen Anne's County is going to be developed whether you like it or not." Everyone knows everyone here, Brad Smith said, and that will be lost if people succumb to the pressure to develop.

Jim Rouse paid a call on the Smiths to explain his Wye Island plan. They had known each other for years. Brad Smith thought Rouse's ideas of limiting the number of boat slips and protecting the shoreline were very forward looking, but he believed Rouse was too idealistic for believing that good development could be encouraged by an example on Wye Island. "The increased population his project will bring in to the Shore will just ruin the county," he said. Smith would control development on the Shore in this way: "You make the rules for developers pretty strict. Developers must provide utilities underground. They must meet county road specifications. You must have very strict rules regarding the drilling of wells and placing of septic tanks. All these requirements on the developer slow him down a great deal." In the abstract that was how Brad Smith would control growth.

For the world he saw from the lawn around his estate, however, Bradford Smith, Jr., wanted growth, along with Rouse, to go away.

———————∿———————

Shepard Krech was raised on Long Island, but he was attracted to the Eastern Shore during his medical schooling at Columbia University. When he and his wife Nora moved to the Shore in 1947, Shep used the Chesapeake Bay ferries to commute between his new medical practice in Talbot County and his patient rounds at the Johns Hopkins University Hospital in Baltimore. He longs for the leisurely days before the building of the bay bridge, when there was always time before the ferry arrived to have a drink and let the exhaustion from night rounds ooze away. He and Nora bought the White House Farm—once one of the many plantations of the Lloyd family—and Shep settled down to a successful medical practice from which he only recently retired. He was chief of staff of the Easton Memorial Hospital. The Krech's large farm extends from Gross Creek to the high bluffs of the Wye East River, and their home overlooks the wooded coves of Wye Island, just a few hundred yards across the river.

One spring afternoon Shep Krech walked me about his farm. The day was warm and comfortable, and so was Shep Krech. He is a handsome, outgoing man, trim and tan, and his blue eyes shine with natural exuberance. Around his wrist he was wearing a slender copper bracelet. The wrist had swollen after Krech tried breaking up three-hundred-pound stones with a sledge hammer for a wall he was building around their summer house in Ireland. His daughter suggested the bracelet. "Pure quackery," he laughed, "but inside of two weeks the arthritis was gone."

The Krechs are enthusiastic wildlife and bird watchers. On the broad lawn that sweeps to the river bluff, they have placed a feeding platform on a post. From their living room facing the river and Wye Island, the Krechs watch cardinals swoop to the tray. After the hunting season, Shep and a radiologist friend would run their boat up and down the Wye East River, picking up crippled geese, setting their wings, and releasing them after they had mended. Once a bald eagle got into the pen and killed two geese. "Nature knows best," said Shep, as we neared a pond from which he guns geese and ducks in the fall and winter. He crouched and with a stick broke apart some dried animal droppings, exposing small whitish flakes that resembled fish scales. They were. We were standing on an otter run. A small branch seemed to be swimming across the pond. "Look at that muskrat," said Shep, "he's taking food to his young." A black duck nested on the pond one spring, and periodically Shep discreetly checked the nest and its eggs. One day the eggs and the duck were gone, and lying next to the nest was a snake's shed skin. "Birds have a tough time with nature," he said. Shep plants honeysuckle bushes ("not the vine") around his farm. In the winter their small berries provide food for twenty covey of bobwhite quail. In the fall Shep likes to take his two German shorthair pointers into the field and hunt a few quail. An unseen bird began making piercing cries. "Do you hear that yellowlegs?" Shep asked. "Isn't that pretty? Just migrating through."

Later that afternoon we walked around the grounds of the neighboring Gross Coate estate with the owner, DeCoursey Tilghman, when Shep suddenly stopped and said excitedly, "Hear that? It's a Baltimore oriole!" My scalp prickled. As a child living in southern California, I had a large book full of pictures of colorful birds that never seemed to get to our neighborhood: redheaded woodpeckers, cardinals, scarlet tanagers, Baltimore orioles. In the illustration the brilliant orange-and-black oriole was as big as a pigeon, so I assumed that they would be easy to locate. To my dismay I have learned that Baltimore orioles are very small and extremely shy. I think they live in Tibet. At 23,000 feet. I raced after Krech around the trees, searching, in vain, high in the branches for the elusive oriole. The oriole, still invisible, called once again from another tree, then apparently flew off to Nepal.

———————cᴧɔ———————

Shep Krech loves to hunt. He has organized the Legal Limit Dove Club, a group of twenty-four friends who each kick in a hundred dollars to rent farmers' fields for pass-shooting at doves. On the Wye East River, below Krech's home, there is a plywood duck blind. Here Shep spent his happiest days, when canvasbacks by the hundreds would roar up the river, flash past the blind, their backs shining, and swing upwind in a figure eight over the decoys. That was great gunning then, he said, and his eyes got that faraway look that Sam Whitby's get when he recalls the gunning days even further back. Shep loves gunning so much that, years ago, he even risked the Stewart wolfhounds, slipping across the river to shoot from an abandoned blind on Wye Island.

One morning Shep and a friend were having a "good shoot" on Wye Island. Two limits of birds lay at their feet, but they kept shooting and stuffing additional canvasbacks in a hollow stump behind the blind. Game wardens watched through binoculars. Shep was fined on the spot, and his shotgun and ducks were seized. Upon the whispered advice of one of the wardens, Shep drove to Easton that afternoon and retrieved his gun and a limit of ducks.

After dinner Shep excused himself and returned to the living room a few moments later cradling a large-scale model of a home. With their children grown, the Krechs found their present home to be too large and costs of upkeep too high, particularly considering that from June to October they stay in Ireland. "We just need a little one-story place with no yard and upkeep," said Nora. As we gathered about the model, Shep lifted off the cardboard roof and revealed four wings, including a guest wing, with fireplaces in the den and dining room and in each of the bedrooms. "And no *grass!*" Shep exclaimed. The site where they planned to build their new home was at the other end of the farm, on Gross Creek; they had already planted pine tree seedlings there to substitute for having to mow a lawn. Across Gross Creek is Gross Coate Farm. The Tilghmans are one of the original Shore families (Tench Tilghman was George Washington's aide de-camp). The Krechs felt confident the Tilghmans would never sell off Gross Coate to developers and ruin the pastoral view from the Krech's new home. For years Shep and Nora had felt secure that Wye Island would remain the same, at least their view of Wye Island. But Rouse had shattered their security.

When, on May 1, 1973, Shep and Nora heard the news that Jim Rouse was planning to buy and develop Wye Island, they were enraged. Shep sat down and wrote Rouse this letter:

Dear Jim:

The proposed development of Wye Island by your company has been received in Talbot County as another potential traumatic insult to the Chesapeake Bay. The Wye East River is shared by both Queen Anne's and Talbot counties, and those of us in Talbot who have a bona fide concern for the future of the Bay are not going to sit back any longer while the Bay is repeatedly raped by a small handful of dollar-greedy people.

. . . Who is going to deliver the health care for several thousand more people suddenly placed in the Memorial Hospital, Easton, service area? Where will you get the additional physicians?. . . Queen Anne's County for years has had insufficient medical coverage to a point where the commissioners have recently come before the Memorial Hospital. . . seeking help in solving this dilemma. Furthermore, it is anticipated that it will take another 8–10 years before the State of Maryland has adequate numbers of physicians and facilities to provide proper health care for all. . . . Couple your proposals with those of the development of Pioneer Point in Queen Anne's County and what a chaotic situation you could produce for the Eastern Shore you profess to love.

No matter how tastefully Wye Island can be developed it can only be deliterious [sic] to the Bay in the long run. There is no such thing as a clean marina. Every boat on the Bay is a potential polluter. Wye Island has a lovely heron rookery, is a resting area for countless thousands of migratory waterfowl. Remove these with "planned development" and who ends up the loser? Are we to say that we humans have more right to Wye Island than the flora and fauna presently there?

I feel very strongly that an indefinite moratorium should be placed on the development of Wye Island. The island is more worthy of its past heritage and future contribution to mankind than to have her fields and woods raped with concrete, bricks, gas and water lines, etc.

Jim, this is only the beginning of the battle to save the Island, but I won't be alone in my feelings, and I shall do all I can with the time at my disposal to prevent your company from carrying out planned murder of the Eastern Shore. I love the Shore. Do you really, Jim? Be happy to discuss all this with you at your convenience. Copies of this letter are going into appropriate hands.

> Sincerely,
> Shep

Shep Krech sent a copy of the letter to Julius Grollman, president of the Queen Anne's County commissioners, with a covering letter urging the commissioners to "have the courage and foresight to deny such development as planned by the Rouse Company." He concluded, "Developing Wye Island in any way other than for total conservation needs would be disastrous to the ecology of the Chesapeake Bay."

Rouse was floored when he received the letter. He had known Shep and Nora Krech for many years. He knew that his proposal would receive some bumpy going. But he had not expected such bitterness. Rouse sat down and wrote this response:

Dear Shep:

Feeling as you do, I am truly grateful that you took the time and trouble to write me, thus giving me the opportunity to respond.

I respect you and Nora and your neighbors on the Talbot County side of Wye Island, as well as those who live on the Queen Anne side and elsewhere in both counties.

We will not rape Wye Island.

We will not murder the Eastern Shore.

We are undertaking to make proposals for Wye Island out of the belief that there are new forms of development for ecologically sensitive land that can respect the land, the water, the fauna and the flora and accommodate rational, sensitive, imaginative development.

I can understand your fears and outrage. I can think of no image in America to which one can point as an adequate demonstration of what ought to be.

Believe me when I say that we are beginning the process in which we are now engaged in the belief that we can identify the possible violations of nature, land and water; set standards to protect against such violations and then follow those standards in the evolution of a plan and a proposal. We believe that such an approach will be radically different from anything you have seen in physical plan and in controls over use of the land and the water by residents of the island.

I meant every word of my letter to the residents of Queen Anne's County, of which I enclose a copy. I thought about it carefully and it describes the responsibility we have assumed as we understand and accept it.

We will be happy to sit down with you and others as our ideas unfold. We will share them with you. We will seek to learn from you and from others in a way that will contribute to the restraints and controls that will shape our ultimate proposals.

Remember, the investment in Wye Island by the present owners makes its continuance as farm land uneconomic and unlikely. Under its present zoning and with traditional development, it can accommodate 400–500 lots with docks at intervals of 200–400 feet around the entire shoreline. This is the old, easy way to rape land and water. By failing to work with us on a better alternative, you may force upon the island the very result you abhor.

You have nothing to lose by keeping your mind open to the possibility that there may be merit in our proposals. I promise you that we are being sincere and responsible in the manner in which they are being developed. Don't pull the trigger until you see your enemy—you might shoot a friend.

I will give you a call on my next trip to the Shore and hope that we can talk about it some more.

Sincerely,
Jim

Jim Rouse soon paid a call on the Krechs. The Rouse Company had just begun its scientific studies of the island and was many months away from a specific plan. Rouse talked of his philosophy about keeping the shoreline from being nibbled to death and about his general hopes of building an island village tailored in every detail to the Eastern Shore. Shep and Nora Krech were not persuaded then, nor were they later, when Jim Rouse showed them the layout of the plan.

———— ⌇ ————

The sun had set, and as we sat in the Krech's spacious living room the cool panorama of the river and Wye Island spread before us in the fading light. A large white bird broke the stillness as it soared around the cove. "Oh, look at that osprey, Shep," Nora said excitedly. We watched the fish hawk flap its way up the river and disappear into the dusk. "Uh-huh," said Shep, "I hope he nests nearby."

"I do too," Nora said.

Shep set the cardboard roof back on their model home and turned to the subject that had come to dominate their thoughts over the last year. "That mock-up looks beautiful," he said, referring to the Rouse scale model. "On paper it all looks good—if he sticks to it." Shep shook his head: "You can't trust him to, though."

If Shep Krech were running the government, he would forbid all developers from touching a blade of marsh grass. Marshes, besides providing habitat and food for aquatic life, help absorb wave velocity and slow erosion. Under the slumping banks of Gross Creek, Shep had dumped oak and beech limbs. Small shoots of marsh grass were already beginning to reestablish in the shallows among the protective limbs. Shep would insist that developers save up to 20 percent of their developments for wild areas and open space. "I'm not against development per se," he said.

But the Rouse Plan for Wye Island, that was different, even though half the island would remain as open space. Shep and Nora looked out at the vanishing trees of darkening Wye Island. Shep Krech's normally calm voice tensed. "By god, it's not that I'm against *houses* on the island!" he said. Then he turned and smiled, quietly saying, "Well. . . I guess I am."

———— ⌀⅓ ————

Says Frank Hardy: "Shep Krech considers Wye Island his private park."

CHAPTER 5

THE SHORE STIFFENS AGAINST
GROWTH

With little difficulty you can find people on the Eastern Shore who will tell you, without a trace of humor in their voices, that they would gladly blow up the Chesapeake Bay Bridge, if that would return the Shore to its former tranquility. A visitor might well conclude that the Shore seems rather tranquil now; and, compared with nearby Washington and Baltimore, it surely is. But to the people who grew up on the Eastern Shore or moved there years ago, the present bears little resemblance to the unhurried days of the ferries.

Not that everything about the past is cherished, for what was pleasurable to some was painful to others. While their children skated on the thick ice that sealed off the entire Chesapeake Bay in the winter of 1936, many parents were desperate, unable to budge their workboats for almost a month. Dick Greaves, then a farmer on Wye Island, figured that when the ice began to break up, Bennett Point would be ground into the bottom of the Wye River and Wye Island would be shoved into Talbot County. The late Walter

Denny, a farmer who once boarded at the Robinson place on Wye Island when he taught school there, also remembered those winters. He slept in the attic. "I like to froze to death," he recalled. "I got so cold one night I went downstairs and got all the papers I could and came back upstairs and stuffed them in between my blankets to keep warm."

When the snows piled up, the roads, which in good weather were bad enough, became impassable. To the increasing frustration and anger of the local people, the state roads commission refused to keep a rotary snowplow on the Shore. Isolation from the Western Shore was fine, but being marooned for days and weeks in your house was something else again.

But for every painful memory, which time has a way of crowding out, there were a dozen pleasant ones, which time embellishes. People on the Shore said: "The city is the place to make one's fortune—the country is a place to live." Before the bay bridge was built, life on the Shore was slower paced and change was less noticeable, principally because it was not accompanied by great numbers of newcomers. Radio gave people on the Shore a contact with the rest of the country without the nuisance of having the rest of the country move in on them. With a battery-powered radio, and a windcharger (which resembled an airplane propeller on a pole) to keep the signal from fading, you could listen to such memorable programs on the Columbia Chain as "The Magic Tenor," "School of the Air," "Current Events," "Skippy," "Kate Smith," and "Morton Downey." When the Clark Motor Company of Hillsboro showed off its new Caterpillar tractor, it attracted the neighboring farmers by offering "Free Talking-Pictures. . . and Free Smokes." A movie then was a major social event, a chance to see friends and visit. The newspapers hyped the movies to seem like a Hollywood personal appearance: "Bing to Sing on Kent Island," "Chevalier Headed for Centre-ville." It was the heyday of baseball. Teams of the Eastern Shore League, like the Kent Island Blue Jays and the Centreville Colts, pulled crowds of three thousand, and the return of the major league's great Jimmy Foxx to his hometown of Sudlersville provided a month of newspaper copy. There were local boxing matches, the circus, and every summer the pre-eminent event of the upper Shore: the Kent Island Fourth of July celebration. It drew eight thousand people in 1936. But the crowds were largely local people. Unlike today, boats that then crowded the landings were mostly local workboats.

The Eastern Shore was never entirely isolated, for steamboats had moved people back and forth across the bay since the nineteenth century. However, few of the people who climbed aboard the steamers at Baltimore's Light Street piers planned to stay on the Shore. Most were going to the beaches at Ocean City, Maryland, and Rehoboth, Delaware. The Eastern Shore was not a place to get to, but to get across. By 1890 rails connected tiny Claiborne (near Saint Michaels) to Ocean City and soon linked Queenstown and Love Point (Kent Island) to the Atlantic Ocean near Rehoboth. For about four dollars you could board the steamer, enjoy a nice breakfast for an additional fifty cents, land in Queenstown, ride the train for two and a half hours, spend the day at the beach, and return the same way to be back in Baltimore by ten at night. If you had to share the steamer home with a few thousand baskets of peaches destined for the Baltimore canneries, the inconvenience was slight. By the 1920s the ferries began replacing the steamers and, in addition to carrying passengers, hauled over their automobiles; the bay-to-ocean passenger rails were then abandoned for roads. The ferries meant some highway congestion—in 1949, 700,000 cars rolled off the ferry at Matapeake on Kent Island—but the people living on the Eastern Shore did not yet feel invaded.

That was all changed when the Chesapeake Bay Bridge opened on July 31, 1952. For half a century there had been talk about bridging the upper bay, a subject that drew heated arguments on the Shore. In the 1930s when the state legislature considered locating the bridge near Baltimore, the Centreville newspaper roared that ". . . Baltimore City is once more trying to foist progress upon the Shore whether we like it or not." Later, when the proposed bridge site was shifted south to connect Annapolis and Kent Island, the paper welcomed the prospect of increased local tax revenues that the more accessible Eastern Shore properties would bring when developed. But many on the Shore, like Jim Rouse's brother, Bill, who led the fight to block the bridge decision, saw only ruination of their way of life. In the middle were the poor—the black oyster shuckers, the women crabmeat pickers, the cannery workers—people who owned little property, if any, who had a faint chance of reaping a bonanza selling land they could never own, and who just hoped that the bay bridge would bring progress to the Shore and some better jobs.

As the bridge dedication ceremonies got under way that broiling summer day in 1952, however, six thousand people stood on Kent Island and simply marveled that this great colossus had actually been built. Almost five years in construction, the bay bridge was built on steel piles driven as deep as two hundred feet into the bottom of the bay, and consisted of 118,000 tons of cement and 60,000 tons of steel, suspended at its highest point two hundred feet over the Baltimore ship channel. Stretching more than four miles over water, it had cost $45 million. The ceremonies began at 10:30 a.m. at Sandy Point and finally ended on Kent Island five and a half blistering hours later. In between, the "Star Spangled Banner" was sung, speeches were given by Governor Theodore R. McKeldin, former governor William Preston Lane, Jr. (for whom the bridge was later named), and many others; a nineteen-gun salute was fired; and the sixty-voice Baltimore & Ohio Railroad Glee Club sang "The Testament of Freedom," accompanied by the U.S. Second Army Band. The governors then got in their Cadillac convertibles and led a stuttering, mile-long car caravan across the bridge to the watermelon stands on the Eastern Shore for another two hours of speeches that ended only when the public address system mercifully broke down. It was a day of ceremonial masochism.

It was also a day of firsts. Francis the Talking Mule, fresh from an appearance at the Democrat's national convention, was the first mule to cross the bridge. The Goodwill Fire Company of Centreville drove the first fire truck across the bridge. Seventy-year-old Alexander B. Townsend, a former native of Centreville, was the first to walk its entire length.

That weekend people jammed the bridge, parking their cars to get out for a view. The snarl of stalled vehicles reached seven miles across Kent Island—an ominous precursor of the gargantuan traffic tie-ups soon to engulf Kent Island and the Western Shore during the summer weekend rush to the Atlantic beaches. Through its three and a quarter centuries of settlement—the tobacco era, the oystering era, the steamboat era—the Eastern Shore had remained relatively unchanged. Now, literally overnight, the Eastern Shore was suddenly thrown open to the automobile and the millions from nearby Washington and Baltimore, many of whom longed for a country place or a summer cottage. And running before them came the subdividers.

The history of places is wrought by many forces, but just as some epidemics can be traced to a single carrier, so, too, are the workings of a single individual often the catalyst of change. When people of the upper Shore talk of the past and future, they speak of but one dividing line, one singular irreversible event: the opening of the Chesapeake Bay Bridge. And in the same breath they usually curse the name of one man: Dave Nichols, member of the state roads commission and the first and biggest land subdivider to hit Kent Island. When Nichols finally went bankrupt in 1961, he left behind thousands of tiny, useless lots, a bewildered county, and the seeds of skepticism about development that would blossom a dozen years later into a countywide subdivision moratorium and a hostility toward growth that would darken Jim Rouse's hopes of saving the Shore from future developers like Nichols.

———— ☙ ————

David M. Nichols was born in Oil City, Pennsylvania. His father ran a roofing company in Asbestos, Maryland, and had been president of the First National Bank in Oil City. But Dave Nichols was fascinated with real estate, and in Baltimore he was soon into the thick of land speculation. He was twice elected as president of the Real Estate Board of Baltimore.

By World War II, Nichols was acquiring properties on the Eastern Shore—stores, gas stations, and other small parcels, mostly in the villages on Kent Island, gradually moving out into the farms. His real estate ads were small at first: "Queenstown—Restaurant, gas station, and apartment, $13,500; Bayfront, 9 rooms, 2 baths, $13,500." By the spring of 1952, with the completion of the bay bridge only months away, Nichols's ads were bigger and louder: "DO YOU HAVE $1,000 and want to buy a real bargain in a home? Well, this is it! Seven rooms and bath. . . plenty of floor sockets. . . compared to other local properties this is worth $10,000. . . the price has been reduced to $7,000. Monthly payments, $55. . . ." Just three months after cars began rolling over the bay bridge, Dave Nichols had already subdivided a number of Kent Island farms into more than 2,800 lots. By 1959, he was hustling land at five subdivisions: Kent Island Estates, Bay City, Harbor View, Romancoke-on-the-Bay, and Cloverfields. Within eight years he had reduced much of Kent Island to 4,681 lots. Nichols was by no means the Shore's first speculative subdivider—he often bought land from the eccentric prison physician from Baltimore, Dr. Theodore Cook, who had been plucking up Kent Island farms since the 1930s—nor would he be the last. But Dave Nichols was surely the largest. And the most notorious.

Nichols was large in many respects; he stood about six-feet-two and weighed well over 350 pounds. He could, and often did, eat three huge steaks at a single sitting. Sleeping during most of the day and working at night, Nichols sent his harassed staff out across the Shore for food at all hours of the morning. An aide gave him rubdowns. Fearing illness, Nichols carried a small drugstore of medicines and pills wherever he went.

Dave Nichols liked operating in style. His aides drove him about the Shore in Cadillacs. In each car Nichols had a mobile phone, and he installed five telephones in his bedroom at Cloverfields. His pilot flew him back and forth between Baltimore and his land office on Kent Island; Nichols also used the amphibious biplane to show Kent Island to prospective buyers.

Dave Nichols used to tell people that he had a vision about cities growing up on Kent Island, and that he would build them. But Nichols built no cities. He built no towns, either. In fact, he built very little, except for a few speculation houses. Dave Nichols was a hustler of lots, not a builder of cities. But he did make one attempt to leave behind his name on one structure; along the highway leading off the bay bridge, Nichols erected Kent Island's first shopping center. By modern standards it was not much of a shopping center—a shoebox-shaped, one-story brick building with Nichols's name chiseled in concrete at one end—but in 1953 the opening of the Nichols Building was a very big event on the Shore. Six thousand orchids were passed out to the ladies, and Jimmie Short and his Silver Saddle Boys gave four performances. Governor McKeldin sent a special greeting, saying, "I like the idea of concentrating these places of business here in a center, rather than attempting to scatter them along our fine new highways." Within a few years, however, the roof fell in. Literally. A heavy snowstorm piled up one winter, and the flat roof of the Nichols Building collapsed into the Acme market.

Although Nichols had been appointed to the state roads commission after he had begun subdividing Kent Island, he was not a bit bashful about using his influence on behalf of his developments. He saw to the paving of the dirt road that led from Love Point down the island past his subdivisions. Soon after the shopping center was built, cars started smashing up on the new dual-lane highway that ran beside it, because Nichols had failed to build an access road to the center. At the urging of Nichols the state roads commission built one, at a cost of $60,000, and paved another road behind the center. At Cloverfields, Nichols entertained politicians and government officials with lavish meals and liquor. He continually nurtured his relationship to Governor McKeldin; he brought the governor over to Cloverfields for his first duck hunt, and gave him a new twelve-gauge automatic shotgun.

During the Kent Island subdivision frenzy, Nichols had proudly stated that he was part of the biggest land boom since the Gold Rush to Sutter's Mill. But more rapidly than the veins had played out in California a century before, Dave Nichols's dreams came to an end. The first signs of trouble—raw sewage—started puddling about the shopping center and adjacent motel within a year of the grand opening. And the state health department found conditions at eleven subdivisions on Kent Island—three of them Nichols's—to be almost as bad. Most of Nichols's subdivisions consisted of tiny plots no wider than fifty feet, and when enthusiastic buyers from Washington started building their summer cottages, they discovered what every farmer on Kent Island knew: the water table was not far beneath the surface of the cornfields. Consequently, septic systems, particularly when jammed close together on small lots, soon spread their putrid contents over the yards. What was first an inconvenience soon became a major health hazard. Although the publicity hardly helped Nichols's lot sales, many of his buyers, having no intention of building a cottage, cared little about adequate percolation. Like Dave Nichols they were speculators, buying cheap and selling high—only on a smaller scale. But if a lot buyer tried to cancel his contract when sewage started running out on the ground, Nichols would threaten with immediate suit.

Nichols did everything to attract buyers, spending a fifth of his paper income on advertising, mostly in the Washington and Baltimore papers and radio stations. Little cash passed in or out of Nichols's pockets, because almost all his transactions were on credit. He borrowed heavily to buy the Kent Island farms and sold lots for credit spread as thin as a crepe. Five dollars down and five dollars a week, the natives remember, giving rise to the common belief that Dave Nichols would sell a lot for "a nickel down and a nickel a week." Thus the opprobrium: "Nichol lots."

But the gold rush to Kent Island never panned out. Nichols had bought and subdivided far more land than he could sell, and many of his buyers backed out of their contracts. With debts of over $1.3 million—eight times his equity in subdivided farms—David M. Nichols filed for bankruptcy in 1961. Five years later he died.

———————— ᴄᴠ᙭ ————————

When the Nichols shopping center opened twenty years ago, it contained seventeen businesses. Now there are six: the Acme market, a clothing store, a pharmacy, a real estate office, a liquor store, and a shop that sells religious books. The sign over the Islander Motel is broken in two and hangs at a drunken angle; the current owner had planned to level the motel and replace it with a McDonald's golden arch, but McDonald's pulled out.

One afternoon I drove down the road past the forlorn memories of the Kent Island land boom—Bay City (which has some houses on the bay, but is hardly a city), Matapeake Estates, Chesapeake Estates, Queen Anne Colony. Near where the road branches off to Romancoke, I turned west toward the bay and drove by a few scattered prefabricated cottages which sat in an open field. On a porch, almost obscured behind a large sign which read Private Property—No Trespassing, stood a large German shepherd. The dog appeared fully prepared to enforce the injunction. At the bay I stopped in front of another, larger sign that read:

THIS COMMUNITY BEACH IS FOR USE OF PROPERTY OWNERS IN KENT ISLAND ESTATES ONLY. . . TRESPASSERS WILL BE PROSECUTED. . . BEACH CLOSES AT 9 P.M. DAILY. . . STATE LAW PROHIBITS FIRES.

The beach was about a hundred yards long and a few yards deep. Weeds grew up in more than occasional patches. A man carefully led his dog around the debris that littered the beach. Pieces of broken sewer pipe had been arranged in makeshift groins, protruding into the water every twenty feet or so in a somewhat futile attempt to arrest the bay's timeless erosion. A few small concrete picnic tables were scattered around. The wooden tops were pulled off most, and the rusted bolts that had once secured them stuck up in the air.

Recently, a group of homeowners in Kent Island Estates tried to get permission from the state to dredge out the silted-in channel of a place called No-Name Creek. The state wildlife biologist expressed concern about the impact which dredging 50,000 cubic yards of muck would have on the clam and oyster beds and other fish and wildlife. The homeowners replied that the clam bed was no longer accessible, the oyster bar had long ago silted over, and the only fish they had seen in the shallow creek were carp. We bought houses here, they said, so that we could use our boats—all we want is to put the creek back in the shape it was when we moved here. When Nichols was selling lots.

Dave Nichols wanted to be liked by the natives, and hoped to be remembered as "the real friend of Kent Island." Many natives of the Shore often preface their comments about Nichols by saying how much they personally liked him. "They'd still all be using outhouses on Kent Island, if it hadn't been for Dave Nichols" and "He was just way ahead of his time."

Yet the final verdict is that Dave Nichols will always be remembered as the man who chopped up Kent Island.

———— ✐ ————

The Chesapeake Bay Bridge ended the relative isolation of the Eastern Shore in a flood of traffic. The travelers to the Atlantic left their imprint passing through. Cars seeped everywhere up and down the back roads of the Shore, carrying many sightseers who had never crossed the bay on the ferries that were now eradicated by the bridge. Just before the bridge opened, the proud town fathers of Centreville had considered putting up a large welcome sign on the edge of town to divert beach travelers through their village. A week after the bridge ceremonies the traffic into Centreville jumped by 50 percent, and the town commissioners pleaded with Dave Nichols to have the state roads commission take over Centreville's main streets and turn them into opposite one-way thoroughfares.

The traffic on summer weekends swelled. By the mid-1960s the people of Queen Anne's County found themselves physically divided into isolated camps. From Friday afternoons until well past midnight, the eastbound lane of U.S. 50—the main highway leading off the bridge and across the Shore to Ocean City—clogged with thousands of sedans, jeeps, campers, and boat-trailers heading for the Atlantic. On Saturday morning, another wave. The chief purpose in building the bay bridge was to give quicker access to the seashore for the millions living in Washington, Annapolis, and Baltimore. This it certainly did, but at a tremendous cost in human frustrations. Those driving to the beach often found themselves jammed up for hours, inching past Annapolis toward the bridge gates. On Sundays, when most Shore people do a little casual driving and visit friends after church, the hordes returned from the beaches and tempers reached the breaking point. The line of cars backed up across Kent Island and sometimes well below Wye Mills, creating a barrier across the county as impregnable as the Pyrenees.

"I don't like the bay bridge," said Howard Melvin. He sat in a swing that hangs from a huge linden tree in his front yard near Wye Mills, and pointed to U.S. 50, a few hundred yards away. "You got to sit there and can't get across the road. In emergencies the Queen Anne's ambulance can't even get across. I've seen the fire engine and the ambulance come over here and sit and blow their whistles for five minutes, and no one will stop to let them through."

Two years ago a second, parallel bay bridge span was completed. On the day it opened, the backup reached seven miles beyond Annapolis; the lights to the bridge approaches had not been properly synchronized. Summer traffic is now smoother, but on many weekends the jam-ups at both ends still occur, and many wonder whether it isn't just a matter of time before the second span will have induced an impenetrable vehicular wall.

———— ❧ ————

For two centuries the population of the upper Shore has remained surprisingly stable. At the time of the Revolutionary War there were about 16,500 people living in Queen Anne's County. In the next hundred years the population undulated between 12,000 and 19,000, leveling off around 14,500 by the Depression and World War II. The 1970 census—eighteen years after the bay bridge—showed a county population of 18,422. In historic perspective, the county does not appear to have been invaded by vast numbers of people. But that is not how the natives viewed the changes wrought by the bridge. They felt overrun by outsiders.

The flood of summer traffic accounted for part of this feeling. But what troubled them most was that Kent Island was changing before their eyes. For years, even centuries, certain farms had been in the same family ownership, and the landscape had remained unchanged. That was all changed by Dave Nichols and the other subdividers of Kent Island and the new city people who followed in their wake. Between 1950–60, when the county population increased by 13 percent, the population of Kent Island jumped by over 40 percent. For Kent Island this constituted traumatic change—and the vibrations rippled out across the upper Shore.

In 1936 the Centreville paper announced that the last farm on Manhattan, a one-acre vegetable plot owned by Joe Benedato, was being turned into an auto trailer camp. Few on the Eastern Shore gave it much notice. Besides, that is New York City, they said; it would never happen here. Thus the subdivision boom of the 1950s came as a rude shock to the people of the Shore. Not every subdivision was crowded with new homes, however. Even as late as 1975, about 80 percent of the lots that Dave Nichols subdivided remain vacant. Of the 6,700 subdivided lots on Kent Island, almost 5,000 have never been built upon. Yet the farms no longer looked the same: the barns were abandoned and the fields in weeds. And the waterfront, where most of the homes were built, was radically changed.

More important, there were now different types of people, strangers, who filtered into the post office to collect their mail and into the general stores to buy their garbage cans.

I don't know anybody in this restaurant. Fifteen years ago that would never have happened. And more people are moving in!

I don't own any of the Eastern Shore. I'm not a squatter, either. But these people come over here and buy a piece of property and pretty soon they're in the church. These people have come here and taken our roots away. They don't like our form of government. Everything we've stood for all these years, they're gonna take it away from us.

When the subdividers rolled over Kent Island, they were able to operate unconstrained by even elementary land use controls. The Eastern Shore is tenaciously conservative, resentful of governmental interference at any level, so there was nothing like a land use plan, much less a zoning or subdivision ordinance, in Queen Anne's County against which Nichols's schemes could be measured. Nichols and the others bought as many farms as their credit allowed and chopped them into however many thousands of little roadless lots that they wished.

To the south, Talbot County was initially spared the worst of the subdivision rush in the 1950s. Being farther from the bay bridge, Talbot was less convenient to Washington and Baltimore. And it had long been settled by richer people, who were not pressed by economic misfortunes to sell out to the developers, and who rallied quickly against any apparent threat to their newfound environment. In 1954, when the Pan American Oil Company tried to locate a refinery along the bay, the gentry of Talbot helped push through the county's first zoning and subdivision ordinance to stop it. Their victory was not without a considerable fight, however; laborers in the county sought the refinery and jobs that they hoped it would bring. And then there was Aldace Freeman Walker, who went on local radio to announce that the zoning ordinance and proposed comprehensive plan were a Communist plot, despotic, and the very essence of totalitarianism. Walker also sued to have the zoning ordinance declared unconstitutional, but the court upheld the ordinance.

The zoning and subdivision ordinances did not completely protect Talbot County from haphazard lot subdivision; they merely assured larger lots and smaller, less-concentrated subdivisions, so that the impact was more spread out geographically and over time.

But in Queen Anne's County, where the evidence of chaos was visible all over Kent Island, the county's first permanent zoning and subdivision law was not enacted until 1965, well after the initial wave of subdividing had passed, leaving behind thousands of weedy vacant lots, overflowing sewage, and strip development along the highway.

Two years before the bay bridge opened, the *Queen Anne's Record–Observer* expressed this somewhat ambivalent concern about the changes that it would bring: "It is not desirable to have a lot of unsightly buildings and shacks spring up to impress the travelers with our lack of modern facilities and planning. The highway scenery should be beautiful and contain restaurants, filling stations and other businesses that would be a credit to our communities." Yet when the state announced its intention to limit access to the new highway cutting across Kent Island from the bridge, the local merchants and farmers—many of whom justifiably worried about access to their severed fields—pressured the governor to back down. The highway was soon bordered with billboards and filling stations.

In 1951, as Dave Nichols planted subdivision stakes around Kent Island Estates, fourteen county businessmen formed a group to push for a zoning ordinance that would keep property values up and protect the roadsides from "unsightly roadhouses, motels, garages and signs." The Record–Observer issued this prophetic plea: "Already on Kent Island summer and year-round developments are mushrooming. Once the bay bridge is opened, this type of building will spring up rapidly and much of it is bound to be undesirable to landowners who own fine waterfront properties or homes if not controlled. . . . Talk of zoning and planning has started in the county, but the time for action is now. Let us lose no time in getting started." For another decade, the most dramatic decade of change in the county's history, no development controls were enacted.

Finally in 1962, Queen Anne's County enacted interim zoning and subdivision ordinances (which finally required the recording of subdivision plats) and appointed a planning commission to administer it. The commission members had no plan to follow, so they operated rather loosely, case by case. But, for the first time in the county, developers now had to meet a few minimal requirements, such as having to ensure that there were roads to the lots they were selling. Julian Tarrant, a planning consultant from Richmond, Virginia, was hired to draw up the county's permanent zoning and subdivision ordinances and draft a comprehensive plan. All were adopted in 1965. The county lacked a full-time staff to administer these new laws, but the earlier permissive attitude toward the subdividers now underwent a skeptical reexamination, reinforced by Nichols's bankruptcy. Hoping to protect the area's rural character and to encourage new development to locate around the small towns, the commissioners zoned about 90 percent of the county for one-acre lots. Wye Island, Bennett Point, most of Wye Neck, and a large stretch along the Chester River, were zoned for five-acre "estate" lots. The plan referred to this 8 percent of the county—the estate zone—as having a ". . . unique value. . . in that [it has] been attracting wealthy and discriminating buyers from other parts of the country who love the beautiful rivers and the congenial atmosphere. . . a real economic advantage to the county."

But not every owner living within the exclusive estate zone agreed with the manner in which the zoning sectioned off the county. One of these was Clarence Miles. Miles lives at White Banks, an estate on the Chester River outside Queenstown. A native of the Shore, he built up an influential law practice in Baltimore, where he helped organize the Greater Baltimore Committee with his close friend Jim Rouse, before moving to White Banks in 1956, where he farms and raises English bulldogs. Miles assisted Arthur Houghton in establishing the Wye Institute and served as its first president. (He would be one of Rouse's few strong supporters for the Wye Island project.) Miles believed that the zoning map had been too hastily and arbitrarily drawn, and he sued the county to have the five-acre limitation struck down. The Maryland high court upheld the ordinance.

To people like Julius Grollman, who had just been elected as a county commissioner, the Miles case was a significant decision, for it gave a legal respectability to the concept of using large-lot zoning as a tool to contain future Dave Nicholses. The zoning of Wye Island for five-acre estates, its first zoning classification, amounted to an almost sacred investiture: once five acres, forever protected. Forever five acres. Jim Rouse would encounter a formidable and rather unbending faith in the sanctity of the five-acre zoning.

———— ✧ ————

Toward the end of the sixties, Queen Anne's County again began to feel the pressure from large-scale developers and subdividers. Now for the first time they moved off Kent Island and along the shoreline of hitherto undeveloped countryside. Among the most dramatic were the series of schemes for Pioneer Point on the Chester River, bringing the specter of change within a few miles of Centreville. It had started in the mid-1960s when a firm called United Nuclear bought the old Raskob estate at Pioneer Point and proposed building a laboratory there. (The late John J. Raskob, former head of General Motors, had bought Pioneer Point before the Depression and built on it no less than two mansions. The stables were paneled like the dens.) The local businessmen, delighted to attract well-paid scientists and engineers, backed a zoning amendment to permit construction of the nuclear laboratory. Brad Smith, however, who was then in the middle of restoring his mansion just across from Pioneer Point, was horrified. He formed the Kent–Queen Anne's Conservation Association and tried unsuccessfully to block the zoning amendment. For unknown reasons, United

Nuclear pulled out of Pioneer Point, opening the way for a man who cast almost as big a shadow as David Nichols.

His name was Charles Rist, a towering, blustering figure (six-feet-seven, weighing three hundred pounds) whose massive recreation subdivision in southern Pennsylvania had made Charnita (*Charles* and *Anita* Rist) a code word for the ultimate in land sales razzle-dazzle. Rist walked into the courthouse at Centreville and threw a big subdivision plat on the table before the planning commission. He announced that he was ready to start pushing lots at Pioneer Point. Look, he told the commission, I'm not in the community-betterment business, I'm a land developer. Just let me in here, and I'll have these lots sold in eighteen months and be out of your hair. The commission summarily turned him down.

Rist decided his project needed a little class. He had a Washington planning firm whip together an ambitious proposal for turning the Raskob estate into a huge convention center, a hotel surrounded by a yacht basin (they called it a "boatel"), a couple of golf courses, a variety of small and large lots, and about 1,100 condominiums. Rist went back to Centreville. He handed his slick brochure to the planning commission members. The cover illustrated a yacht as big as a navy cruiser slicing through the water. At the bow stood a man in a yachting blazer and cap, holding a spyglass to his eye. In the background, a twin-engine seaplane appeared to be crash-diving into the pyramidical boatel. The brochure talked of polo for the discriminating equestrian and the posh exclusivity that Rist would bring to Pioneer Point: "It is planned that the yachtsman will require a more casual atmosphere while near his boat, and therefore should be separated from the more restrictive hotel. Evening activities though would be more formal. . . consequently he would arrive at the hotel in formal attire via the hotel transportation system." Rist said he planned to retire and live at Pioneer Point. Even to the untutored, Charles Rist's flamboyant proposal had a tinny ring, but to county officials now thoroughly shell-shocked from the twenty-year siege over Kent Island, this white-tie-and-tails presentation came as a bad joke. The planning commission told Rist that it would take about three years for them to assess the impact of his proposed development upon the county. Until then, well, he would just have to wait.

But Rist could not wait—he was going broke. Charnita was beginning to unravel. The Commonwealth of Pennsylvania had slapped a construction moratorium on most of his lots after finding them unsuitable for septic tanks. Lot owners and citizen groups filed charges of fraud and misrepresentation against Charnita. To get money Rist had to move some property fast. He found it in Washington, in a Victorian building on Sixteenth Street that had almost as many antennae on its roof as the U.S.S. *Forrestal:* the embassy of the Union of Soviet Socialist Republics, the biggest landowner in the world. For a number of summers the Russians had rented places on the Eastern Shore for their vacationing embassy staff and families—for two years they rented the Duck House on Wye Island from the Hardys but they wanted something more substantial and permanent. When they saw the elegant Raskob mansions, the swimming pool, tennis courts, and guest houses at Pioneer Point, the Russians clapped their hands, and Rist carved out for them forty-five adjoining acres for a little under two million dollars.

But the Russian sale at Pioneer Point could not save Charnita's collapsing finances. Rist filed for bankruptcy. (In 1975 he was shot and killed at his emerald mine in North Carolina.) Pioneer Point was sold to a Fort Lauderdale developer by the name of Warren A. McFadden and his New York partner Edward Halloran. McFadden and Halloran hired Julian Tarrant, the Richmond planner who had drafted the county's comprehensive plan and land use ordinances, to prepare a plan for developing Pioneer Point into two villages, town houses, scattered dwellings, and a golf course. The plan required a zoning amendment to permit clustering of apartments and town houses. The planning commission unanimously recommended disapproval. Undeterred, McFadden went before the county commissioners on the afternoon of August 29, 1973, to plead for the zoning amendment. Because of the big turnout, the commissioners held the meeting upstairs in the more spacious courtroom. It was the same summer that the Rouse people were making their measurements of the Wye River and debating their many alternative plans for Wye Island.

The witnesses for the developers extolled the virtues of planned unit development and clustering. Bobby Price, attorney for the planning commission and one of the more influential county officials, could contain himself no longer. Price's home on Reed's Creek backs right up to Pioneer Point; he did not relish the thought of seeing the farmland behind his house turned into condominiums. He stood up and said that in the past several years at least sixteen other developers had been before the county officials to explain the differences between cluster and conventional development with graphs and charts, and that they did not need another lecture on the subject. Sitting in the courtroom that afternoon to get a feel for the commissioners was Scott Ditch of the Rouse Company, and Ditch wrote this down in his notebook. Julius Grollman then stretched out his feet and squinted at the developers. Y'know, he said, if we require developers to cluster their houses and leave a lot of open space, it will only attract a lot of outsiders over here to live on the Shore—better we stick to our ordinary subdivision grids and discourage people from moving over here. Ditch wrote this down, too.

Julian Tarrant gamely argued that a zoning amendment to allow clustering would simply be a rational acceptance of change. He was unprepared for the question that followed.

"Yeh, in other words, you do like change," said a local realtor, "like you said in the publications and professional magazines and everything. You like that, huh?"

"Well," said Tarrant, "like someone said, the state of the art advances. If it didn't, it would stagnate."

"Well, do you like the women in *Vogue* magazine?" the realtor asked. Julian Tarrant looked at the man blankly. "I don't read *Vogue* magazine," he replied cautiously.

"Well, I do," the realtor declared, "because my wife's mother gave it to us, and believe me, if that's change, they leave a lot to be desired."

The zoning amendment was unanimously rejected by the county commissioners.

Looming over the county now were five major developments: Pioneer Point, Prospect Plantation on Piney Neck, the Myander farm subdivision adjacent to Kent Island Estates, a proposal for high-rises next to the bay bridge, and the yet-unannounced Rouse plan for Wye Island. The county officials felt they were being overwhelmed. On November 1, 1973, while the Rouse Company was still shaping its plan for Wye Island, Queen Anne's County declared a moratorium on the approval of any subdivisions of more than five lots until completion of a new county plan and zoning and subdivision ordinances.

——————— ᴄᴧᴐ ———————

When they enacted the subdivision moratorium, the Queen Anne's County commissioners decided it was time to hire a full-time professional

WYE ISLAND

staff. They chose Col. George Aldridge, Jr., then retiring from a distinguished army career (recipient of, among other medals, the Silver Star, the Bronze Star, and the Distinguished Flying Cross with five oak-leaf clusters, all awarded for heroism under fire). A native of the county, Aldridge was hired as the county's first zoning official and was soon made the first county administrator. Robin Wood, a young planner and also a native of the county came in to manage the planning and zoning functions. The first public works director was hired, a professional engineer by the name of Andy Bristow. Aldridge quickly put his growing staff into committee formation and gave them military-sounding names. One he called the STAC—Staff Technical Advisory Committee. It consisted of the staff members with responsibilities related to land development. The STAC met each week to go over the details of pending subdivision proposals. Although the moratorium temporarily suspended the approval of most new subdivision plats, Aldridge often used the STAC meetings to sound out those developers who were waiting in the wings for the ban to be lifted.

To one such meeting, about a month before Rouse unveiled his plans in the nearby library, came Harold Greenberg to chat about his plans to subdivide the old Mylander farm on Kent Island. With him Greenberg brought his partner Arnold Wolfe, their Centreville attorney John Sause, and their planning consultant from Ocean City, Ruffin Maddox. The four sat on one side of the courthouse room that serves jointly as the meeting place for the county commissioners and the planning commission, as well as George Aldridge's office. The county staff sat in a semicircle; they wore work boots and flannel shirts, except for Wood and Aldridge who both wore ties. Greenberg, Wolfe, and Maddox made a visible contrast: new dark suits, manicured nails, razor cut hair, shiny crinkled loafers, and constant smiles. Sause stuck his feet on a table and tilted far back in his chair. He didn't smile. He clenched a pipe between his teeth and often gave the appearance of being asleep.

Maddox spread a subdivision map on the table before the staff: a 300-acre farm to become a 200-boat marina, a pool, tennis courts, and 285 lots, ranging in size from a half-acre to an acre. How do you plan to treat the sewage? Wood asked. Each buyer would have to put in his own septic tank, Maddox replied, smiling. Sorry, said Wood, our subdivision ordinance forbids individual septic tanks on subdivisions of more than 99 lots—you'll need central sewer and water. There was a long silence. Maddox glanced over at Sause, who seemed to be resting his eyes. George Aldridge took a call on the phone. Look, said Maddox, still smiling, we are here to please, and we know that the soils may not be the best for septic tanks, but what difference should it make how many total septic tanks you have, provided that each of the lots is big enough and percs? Before Robin Wood could respond, John Sause opened his eyes, squinted between his crossed feet at Wood, and said he had a proposition to make: Why not run percolation tests on all the lots, and if they perc, the county, after footing the bill, could grant my clients a variance from the requirement for central sewerage? George Aldridge hung up the phone and turned to Sause. Aldridge debates with few words. No, he said, it is a county ordinance, and I am going to enforce it.

Harold Greenberg stood up, smiling, and told of the dreamy summers he had spent coming over to Kent Island from his home near Washington to fish. He related how much he cared about the Eastern Shore and how, although he was a "Western Shore developer" (Greenberg had never before subdivided land outside of the Washington suburbs), he always endeavored to "improve" a piece of land. He had selected the Mylander farm because it simply looked like a good piece to develop. He said, well, I'm not really a developer in the true sense, because I do not contemplate building any houses; for that matter, most of the lot buyers will be purchasing land as a "holding proposition." When they build their houses, if ever, it will be much later—say, in fifteen or twenty years.

Harold Greenberg kept smiling at the unsmiling county staff and gave a little laugh, "What do I do with the balance of the land if I can only develop ninety-nine lots?"

George Aldridge looked up from some papers he had been reading and responded, "What name do you intend giving this development?"

Greenberg's eyes brightened. "I can remember when I first visited Kent Island, and I thought what a paradise the Eastern Shore was. . . so I would like to call our development Paradise Island. . . of course, if there is any objection, I would be only too glad to change it, because I really only pulled the name out of a hat. I intend to retire and live there myself."

Aldridge looked silently and expressionlessly at Greenberg.

To no one in particular, Greenberg mused about the future of Paradise Island if he were limited to ninety-nine lots; he said nothing about financing a central sewer and water facility. He was finally answered by his attorney, John Sause, who exploded from his seeming slumber: "The county just doesn't *want* any more subdivisions!" Sause continued to snarl about the county's "unconstitutional growth limit" and how a federal district court in California had so ruled, in the town of, the town of. . . Sause stammered, trying to remember the case. "Are you referring to Petaluma?" George Aldridge quietly asked.

The meeting ended with Harold Greenberg saying, in a tone of mild desperation, that he would be unable to make the "entire piece pay out" if be could not subdivide and sell more than ninety-nine lots because of the septic tank limitation. "I will have to consider selling the rest of the farm to another developer," he said, continuing to smile. Robin Wood smiled and thanked him.

Two months later, when Greenberg tried to get state permission to dredge and fill part of the marsh at Paradise Island for his proposed marina, the hearing room was filled with people from adjacent Kent Island Estates and Queen Anne's Colony who loudly objected about the potential increase in boat traffic. Among those homeowners were the developer of Queen Anne's Colony and another developer of a subdivision farther down the island.

The county had finally stiffened against development.

———— ⌒⌒ ————

Julius Grollman has lived all his life on Kent Island, over the family store in Stevensville. The original Grollman store, which his parents opened seventy years ago when they immigrated here from Lithuania, is down the street. Grollman's father began his career in America by walking from farm to farm on Kent Island, peddling kitchen utensils from a pack he carried on his back. Stevensville is a small village just off the main highway near the bay bridge. Its center is literally the Grollman store—the road through town, a distance of not much more than a block, doglegs around it. People in Stevensville call it "the liquor store" because the Grollmans sell liquor, displayed on seven long shelves behind the counter, but the Grollmans also sell beer and soft drinks, which are crated on the floor or kept chilled in a red Coca-Cola cooler. In fact, there is very little that Julius Grollman does not sell in this eclectic mercantile museum, where over the years items have been left on the floor, stuffed into shelves, shoved into corners, or piled upon other boxes in no particular order. A casual sample: work boots, oxfords, tennis shoes, coveralls, Schweppes tonic, rubber boots made in Japan, an oil heater, boxes of nails and spikes, shovels, crab nets, rakes, hand tongs, light bulbs, pellets for water softeners, hasps, hinges, toggle bolts, caps, thimbles, a case full of thread spools, rolls of nylon and manila rope, orange life jackets, kerosene lamps, straw hats, mole and mousetraps, plastic garbage cans, charcoal briquettes, kerosene stoves, hammers, hatchets, wrenches, barrel bolts, folding wood rules, panty hose, long underwear, razor blades, an anchor, sections of stove pipe, dark glasses, metal lunch boxes, washtubs, galvanized crab pots, coal scuttles, mason jar caps, hog rings, copper wire, mailboxes, blankets, sleeping bags, and whisk brooms. I once asked Julius Grollman whether there were any items he did not carry in the store: "I don't sell radios," he said.

Except for Tuesdays, which he spends in Centreville at county commissioners' meetings, Julius Grollman can usually be found at the store supine in a swivel chair, his feet on the rolltop desk that serves as part of the counter. Grollman has no hired help—his wife and teenage children share the duties at the store. "When you hire someone, with all that paperwork," he said to me one morning from his relaxed position behind the desk, "it isn't worth it." A niece who took art in college painted a still life of the liquor shelves behind the counter. Grollman hung it on the wall next to the desk. Suspended directly over one end of the counter was a large air-filled plastic tomato, about the size of a foot cushion, advertising Seagram's Bloody Marys.

The door slammed shut and a pinched man with skin practically blackened by the sun shuffled in. The man reached into the pocket of his grease stained, quilted jacket, pulled out a few crumpled bills, and pressed them onto the counter. Not a man of sudden movements, Julius Grollman slowly uncrossed his ankles and dragged his legs off the littered desk. From one of the shelves he pulled down a half-pint of Four Roses. He wrapped it in a small paper bag and handed it to the man. The man shuffled out of the store, the door banging behind him. Not a single word between either man had embroidered the transaction. Julius Grollman dropped back into the swivel chair and stretched out again.

Julius Grollman has watched change come to his native Kent Island— first, the few summer cottages, then the bridge and Dave Nichols, bringing traffic and subdivision flags. "Dave Nichols was a real nice fella," he said slowly, "but he really screwed up Kent Island. Took a lot of farmland out of production. It's something the county will have to live with for a long time. . . maybe forever."

The door banged again. "Got my ticket?" a loud voice asked, and a swarthy man rocked like a bear up to the counter. He peered at a group of numbers posted above it. Grollman is one of hundreds of local merchants in Maryland who sell the state lottery tickets. In his shirt pocket he keeps an assortment of ballpoint pens and mechanical pencils clipped to a plastic holder, which displays a wishbone and the words: BUY NOW. . . TWIN WIN. For each weekly lottery, Julius Grollman buys about four hundred tickets and gets five cents from the state for every dollar's worth of tickets he sells. Those left over, about a hundred, are his for the drawing. Julius Grollman has never sold himself a winner, however. To another customer who came in and bought a fistful of lottery tickets, Julius Grollman told about a garbage truck driver who won $50,000 the previous week.

"Black or white man?" the customer asked.

"White," said Grollman.

Newcomers wander through Grollman's store as if it were a strange antique shop, poking blindly here and there among the treasures. Over the years the natives have acquired a feel for the store and a working knowledge of where in it they would most likely find what they are looking for; their transactions are accomplished with a minimum of conversation. Late in the morning, a man with a crew haircut entered the store and walked directly to a large spool of black electrician's wire, which he immediately began wrapping in tight coils between his palm and brown muscled arm. All he said to Grollman was that he needed twenty-five feet of cord.

"Lookwise or measurewise?" asked Grollman.

The man made a few more wraps, then he stopped and held his hand at the spot where he wanted Julius Grollman to cut. Grollman brought over a yardstick and measured off the amount of wire that the man had looped off the spool. It was exactly twenty five feet to where the man grasped the wire.

Government is still a rather informal business in Queen Anne's County, and occasionally some of it is transacted in Grollman's store. The door opened to admit a portly man in a white T-shirt streaked with sweat and grease. He told Grollman that he had just finished mowing the shoulders of the county road at Cloverfields. Cloverfields, another Nichols legacy—972 lots, 907 still vacant. "Just bring me down your time, and I'll sign your slip," Grollman told him. The man nodded his head and left. This bit of official business over, Julius Grollman settled back again in his swivel chair, gathered up his legs and laid them over the cash register, shoved his glasses back on his forehead, and yawned. Julius Grollman yawns quite often.

What did he think of the Rouse proposal, I asked him. "I don't own a boat myself, but I wouldn't buy a lot that said I couldn't have a boat," he said. Of Rouse's plans to require all island boats to have holding tanks for disposal into a receptacle at the village dock, Julius Grollman was unimpressed. "They'll just dump those holding tanks in the bay. I've heard it said that some of these people with house trailers will just pull off on a side road and dump their tanks. And there's no way you could require a man to lock his head and enforce it."

Julius Grollman pulled off his glasses, rubbed his eyes and yawned again. "I wouldn't want to see anything under five acres on Wye Island." He settled lower in the chair, which now appeared to be precariously tipped. "The less people on Wye Island the better," he said.

CHAPTER 6

OUTSIDERS AND INSIDERS

Amonth after Jim Rouse presented his plan in the Queen Anne's County library, Dickson Preston drove over the narrows bridge and down Wye Island. It was a warm spring day, a cloudless sky of deep blue. Most of the wintering geese had returned to their nesting grounds in Canada, but hundreds still strutted among Arthur Houghton's Aberdeen Angus, snipping the pastures of Wye Plantation. The Bennett Point–Piney Neck Citizens Association had asked Preston to consider organizing their public relations campaign to counteract the Rouse Company project. On this day Preston was poking around the island, hoping to resolve his own ambivalence about its future. Although Preston is an active conservationist, he dislikes elitism, and he had sensed its presence around Wye Island's wealthy fringes. Also, he had misgivings about becoming a public-relations flack; he would rather write history, which is less a hobby than it is his second profession. About ten years ago Preston retired as White House correspondent for the Scripps–Howard Newspapers, bought a house and five acres on the water, and began writing local history articles for the Easton paper.

From the roof of Gordon Shawn's deserted farmhouse two buzzards watched Preston drive by. He passed through the cool Shawn woods, and down the tunnel of towering hedgerows toward Bordley Point. Preston stopped the car at the base of a massive dead tree and got out. His eyes widened. Preston is fascinated by big trees. In Wye Mills, near where the Whitbys live, there grows a mammoth oak, the largest white oak in the United States, which began as a seedling over 430 years ago, when Henry VIII was cutting his way through matrimony. Preston had written a book of the Wye Oak's rich history (*Wye Oak—The History of a Great Tree*, Cambridge, Maryland: Tidewater Publishers, 1962). Preston enthusiastically tromped through the underbrush around the tree and dictated his observations into a cassette tape recorder. Years ago Howard Melvin had grown up on this farm; some of Jacqueline Stewart's water towers stood nearby, almost obscured in the verdure. Preston wanted to get the state forester out to the island to measure the tree. "I think the trunk is thicker than the Wye Oak," he said. "Now, if I could just find out that Rouse is going to drown this tree, we could beat him," he laughed facetiously. "That's how we saved the Tuckahoe!" (The Tuckahoe is a swampy stream that forms much of the meandering eastern boundary of Queen Anne's and Talbot counties. When the state of Maryland proposed damming it for a state park of large lakes, Preston and other conservationists blocked the big lake scheme by publicizing the fact that it would flood and thus drown a huge swamp oak that grew in the boggy floodplain. The American Forestry Association later certified it to be the largest swamp oak in the United States.)

When Preston drove back to the Wye Narrows bridge, he wondered aloud about the wisdom of opposing the Rouse plan, and he was decidedly uneasy. (Later Preston told me that had the PR job been firmly offered him, he would have turned it down.) He was not sure whether it was worth saving Wye Island for "those multimillionaires." Besides, he said, if the island were going to be developed anyway, maybe Rouse's village and large estates would be better than a house and pier on every five acres.

Carl Blakely, on the other hand, believes that five acres is just right. Ten years ago he moved to the Shore and built a house along the Wye River above Bennett Point. Every morning he drives an hour and a quarter to his office in Washington, DC. On the job, Carl Blakely is director of procurement for the Federal Energy Administration. On the Shore, he is president of the Bennett Point–Piney Neck Citizens Association, which he helped form to combat nearby Prospect Plantation. There, a developer on Eastern Bay was subdividing 324 one-acre lots and a smaller number of five-acre lots, to which canals would be dug to create artificial waterfronts and boating access. Blakely did not approve of the one-acre lots and the canals because he felt that they would attract too many people and too many boats. But five-acre parcels he could accept: you get "the right kind of people"—and not too many—with five-acre lots, he said. That's good land planning. What Blakely disliked was cluster housing. Like Rouse's Wye Island village. To Carl Blakely, cluster housing meant younger people crowding into marinas and families with kids in school, and that implied more taxes for schools and fire protection and police and all that. "If a fellow has five acres, there is just so much land in the county, and the protection to us is that we can know absolutely how many people there will be," he said. Blakely has nothing against people with children, but he wants to see them on five acres and in single-family houses. Not in cluster dwellings. Not like Columbia. Not on Wye Island.

———————— ⌀ ————————

About three weeks after Dickson Preston clattered off the narrows bridge, the Bennett Point–Piney Neck Citizens Association—about fifty or sixty well-dressed men and women, most of them retired and in their sixties—gathered in the small auditorium of the Grasonville Intermediate School to discuss their Wye Island strategy. Frank Hardy sat, arms folded, in the back row with Bill Chaires, his land-sales manager. Blakely was a little uncomfortable having Hardy as a member of the association—at the last meeting Frank Hardy had presented details of the Rouse plan for Wye Island. Brad Smith chatted with an acquaintance. Carl Blakely had invited Smith, hoping to attract funds from his "upcounty" citizens group for the PR campaign against Rouse. Blakely and the other officers of the association took their seats in front of the stage, backdropped by an avocado curtain. Children's paintings of lollipop trees on construction paper were stuck about the tile walls with masking tape. The treasurer reported: $695.01 in cash on hand.

Carl Blakely made a short welcoming speech, saying he was "very heart-
ened, personally" to see such solidarity among the citizens of the county
in their growing opposition to new large-scale developments. Frank Hardy
glared. Blakely then introduced the special guests: Senator John Miller,
who represents the counties of the upper Eastern Shore in the state senate, and Carter Hickman, seasoned member of the House of Delegates for
Queen Anne's County. Hickman, a distinguished-looking man with white
hair spoke first. Other than Rouse's plan, he said, there are about four options for how Wye Island could be handled. The federal government could
buy the island and make it a wildlife refuge. A number of heads nodded
in agreement. Or, Hickman said, "And this is the option the Open Spaces
Committee on which I serve favors," the state of Maryland could buy ease-
ments on the island for the difference between its development value and
its value as farmland, and then allow someone to operate the island as a
farm. Of course, he said, the island could be developed into five-acre lots as
the county zoning provides, "but that would, in my opinion, be worse than
the Rouse plan." The last option, Hickman continued, was for Maryland to
buy Wye Island for a state park. At this, the members of the Bennett Point–
Piney Neck Citizens Association jerked as if they had touched a shorted
wire. Over the rumbling discord Carter Hickman quickly added, "And I can
assure you *that* won't be done!"

Senator Miller stood up and spread out his hands: "I don't think you
want Wye Island as a state park, 'cause it would bring in a class of people
you wouldn't want. I can assure you, they wouldn't be from Queen Anne's
and Talbot counties," he said to the audience of mostly recent arrivals from
cities elsewhere.

Edgar Bryan looked around the room at tailored women in cashmere
sweaters and distinguished men in sports coats and L. L. Bean gum shoes.
Bryan did not see any natives. He was within walking distance of the farm
he had lived on for eighty-six years—surrounded by strangers. Drawing
his slight frame to his feet, Edgar Bryan asked, "How binding is an ease-
ment? You get a new bunch of county commissioners fifteen years from
now and. . . ."

It would be in perpetuity, forever, Hickman answered.

Edgar Bryan frowned. "You mean a man would have a piece of property
and two or three children, and he couldn't build a house?"

That would have to be worked out in the contract, Miller said.

A big, rangy man by the name of Jerome Gebhardt stood up and said he agreed that a state park would not be desirable. He said that five-acre zoning on Wye Island was not so good, either. He continued, "I'm for open space, wildlife, and such." Gebhardt had recently retired and moved into a house on the Wye River, directly across from Wye Island. He had been appointed by the association to seek out a public relations man to counteract Rouse. He was the one who had sounded out Dickson Preston.

"The ideal thing would be to keep Wye Island exactly as it is," said Senator Miller.

A number of people asked Miller and Hickman about the recently passed Maryland land use law. The bill initially had been drafted to give the state certain powers over local governments in controlling development in "areas of critical environmental concern." The Eastern Shore's influence in the state legislature has eroded more than its shoreline, but the specter of the "Governor Mandel–Baltimore–Western Shore crowd" reaching into the heart of local government was enough to send the fiercely independent Eastern Shore into near apoplexy: "They'll try and turn the whole Shore into a 'critical area!'" Hickman, Miller, and the entire Eastern Shore delegation had fought the bill and succeeded in reducing the state's authority to little more than an advisory role.

Frank Hardy stood up slowly. When there was a pause in the discussion, he announced that he wanted to say something. Hardy's voice is deep and resonant; the more people he is addressing, the deeper and more authoritative it becomes. "If you rely on the Open Space Program and the Department of Natural Resources, I suggest you get legal counsel, because under the Open Space Program no money can be used for acquisition of land if access is then denied to the public." His voice was growing louder and descending in pitch. "And then you've got your park," he boomed. Sentence pronounced, Hardy's dark eyes narrowed and through tense jaws, his face seeming to give off heat, he said, "And *that* I'll fight as hard as anyone in this room! And I say that as one of the owners of Wye Island." Frank Hardy sat down, confident and defiant. For a moment no one said anything.

What was running through the minds of everyone in the Grasonville Intermediate School auditorium that night was this: God help us if Wye Island ever becomes a *park!*

On the same day that the Rouse Company announced it had acquired options to develop Wye Island, Thomas Hunter Lowe, then speaker of the Maryland House of Delegates and a native of the Eastern Shore, publicly recommended to Gov. Marvin Mandel that the state of Maryland buy Wye Island and make it a state park. For all the support his proposal got on the Shore, Lowe might just as well have recommended that the state prison be relocated on Wye Island, or the state establish a leper colony there, or the Defense Department use it for testing and storing nerve gas.

To most natives and the newcomers alike, a public park—or a public landing, or a public anything—seems to be the ultimate debasement of all they hold dear about the Eastern Shore. It means outsiders. It means riffraff from the city. It means trash, crowding, campers, and boat trailers—chickenneckers. It means blacks from Baltimore.

"If Rouse don't develop Wye Island," a native of Wye Island once said, "the state will turn it into a park and it will be full of coloreds. *Free!*" He spat the word out. "Anything that's free is no good."

Another saw the same threat if Rouse *did* develop the island. "The goddamn liberals. . . the goddamn Democrats. . . they'll choke your life out for the goddamn niggers. Rouse says he's gonna put a golf course on Wye Island. But not for the people who were born here. For a bunch of niggers who he'll bring down!"

There are virtually no parks in either Queen Anne's or Talbot counties, other than the recently opened Tuckahoe. But people on the Shore have seen parks; many will tell you, "Go back across the bay bridge and look at that beach on the Western Shore." The place they are referring to is Sandy Point State Park, once a busy terminal for the now defunct bay ferry service, now a public beach on the edge of Annapolis next to the bay bridge approach. The people who stretch out on the beach there are black. "See all them niggers? *That's* what a park is!"

There is a saying among the natives, proudly echoed by those from the city who thrill to live on land that must have had a romantic past, that "the Eastern Shore was founded on tobacco and tolerance." About tobacco there is little dispute, but as to the system of slavery that pervaded Shore society for more than two centuries, there is ample room to debate how well it was characterized by tolerance.

Many books about the history of the Eastern Shore dwell on the elegance of the old manor plantations, the exquisite taste and the wealth of the landed families, how they shipped in their silver goblets, their lace and embroidered linen from England and Ireland, and their madeira from Spain—and how decently they treated their slaves. The Land of Pleasant Living. But there was one man who wrote about slavery, particularly on the Eastern Shore of Maryland where he was born and raised, who did not agree. He was Frederick Douglass, and he wrote this about his birthplace:

In Talbot county, Eastern Shore, Maryland, near Easton, there is a small district of country, thinly populated, and remarkable for nothing that I know of more than for the worn-out, sandy, desertlike appearance of its soil, the general dilapidation of its farms and fences, the indigent and spiritless character of its inhabitants, and the prevalence of ague and fever. The name of this singularly unpromising and truly famine-stricken district is Tuckahoe, a name well known to all Marylanders, black and white. It was given to this section of country probably, at the first, merely in derision. . . that name has stuck. . . and it is seldom mentioned but with contempt and derision, on account of the barrenness of its soil, and the ignorance, indolence and poverty of its people. Decay and ruin are everywhere visible, and the thin population of the place would have quitted it long ago, but for the Choptank river, which runs through it, from which they take abundance of shad and herring, and plenty of ague and fever.

It was on the Tuckahoe that Frederick Douglass was born a slave in 1817. He lived with his grandparents in a shack of clay, wood, and straw. Douglass often heard them refer to "Old Master," but he was too young to know or care who Old Master was—the man who owned the land they lived on, who owned his grandmother (his grandfather was a freeman), and who owned Frederick Douglass.

When Douglass reached seven years of age, Old Master, one Aaron Anthony, ordered him brought to live on the huge Lloyd plantation of Wye House, across from Wye Island. Anthony served there as manager for the vast estate of Edward Lloyd V. Douglass's grandmother walked and carried him the dozen miles or so across Talbot County from the Tuckahoe to the Wye. As Freddie stood by the slave quarters at Wye House and watched a group of children play (among them a brother and two sisters, whom he had never met), his grandmother quietly slipped away into the woods. Douglass never saw her again.

Slave children were customarily separated from their parents at a very early age, so it was not until Douglass moved to the Lloyd plantation that he first saw his mother, and then only sporadically (he never knew his father). His mother was a field hand on another plantation a dozen miles away. A few evenings she got away unnoticed, and walked in darkness, exhausted and weary, through woods and brambles to the Wye. She would find her frightened son on the floor of a cramped closet, tucked into a cornmeal sack, and would hold him through the night. Well before daybreak, while Douglass slept, she would return to the distant fields. Douglass's mother soon grew ill, but her children were not allowed to visit her. She died alone.

Douglass would later write (1855) in one of his autobiographies, *My Bondage and My Freedom*, that although it was believed customary for slaveholders to give their slaves enough to eat, the food was usually unpalatable and the Lloyd plantation, and others he lived on, were places of constant hunger. As a child, Frederick Douglass got only cornmeal mush, which was dumped into a wooden trough and placed on the kitchen floor or the ground outside. The slave children were forced to fight among themselves for as much as they could scoop up with oyster shells and pieces of shingle.

Douglass dreaded the cold of winter even more than he did the gnawing hunger, for his clothes consisted of a coarse sackcloth of tow-linen. He had no pants, no jacket, no shoes nor socks, and the cracks in his feet split wide enough to lay a pen in. For warmth during the day Douglass sought out the sunny side of the house, and when storms drove him indoors he crawled into the kitchen fireplace among the warm ashes.

Douglass learned why slaves always sang in the fields (so the overseers would know they were there and working), why slaves always assured others of the kindness of their own masters and how contented and well-fed they were (the slave traders and inescapable Georgia fields awaited slaves who said otherwise), and why a black man had to tip his hat and step aside to let white people pass (anything less than subservient behavior meant a whipping, or worse).

Slavery was a life of absolutes. By the overseer's slightest whim a slave could be torn apart by the cowskin, which the overseer always carried. The cowskin was a three-foot strip of dried ox hide, about an inch thick at the handle, and tapering to a fine point at the end. Douglass considered it more deadly than the cat-o-ninetails. "It condenses the whole strength of the arm to a single point," he wrote, "and comes with a spring that makes the air whistle. It is a terrible instrument, and is so handy, that the overseer can always have it on his person, and ready for use. The temptation to use it is ever strong, and an overseer can, if disposed, always have cause for using it. With him, it is literally a word and a blow, and, in most cases, the blow comes first." Through a hole in his closet Douglass watched his slave master tie a slave woman by the wrists to a ceiling joist and then pulp her back with the cowskin. It was a sight Douglass would witness and be victim of many times, as he discovered that violence was a fact of life on the plantations. And a fact of death. It was "worth but half a cent to kill a nigger and half a cent to bury him," went a common saying on the Eastern Shore. Douglass saw Lloyd's overseer drive a slave into the Wye with his cowskin and then blow the man's head off with a pistol; one of Bordley's descendants on Wye Island shot an elderly slave in the back while the old man gathered oysters, as most slaves did to supplement their meager diets.

After about three years on the Lloyd plantation, Frederick Douglass was shipped off to Baltimore to stay with his new slave master's brother, and there, for the next seven years, Douglass lived in relatively humane conditions. The mistress of the house made the error of teaching him the alphabet, opening Douglass's searching mind to the world and the power of the English word. He read a schoolbook entitled *The Columbian Orator* and was thrilled by the speeches of Pitt and Fox. Douglass got his hands on *The Baltimore American*, the city's newspaper, and read with fascination about Nat Turner's insurrection. He discovered that there were slaves who had run away to freedom, and that there was a movement to aid them and abolish slavery. He read Patrick Henry's speech about liberty or death.

When Douglass returned to the Eastern Shore, to Saint Michaels, in 1833, he was a determined abolitionist. But he was still a slave, and many years were to elapse before he broke free.

After one of his escape plans was discovered, Douglass and his co-conspirators were dragged barefoot in irons fifteen miles to the Easton jail. When Frederick Douglass was sent back to Baltimore to work as a ship's caulker, he finally made his permanent escape.

———— ∽ ————

Such was the tenacious hold of slavery on the Eastern Shore—and the white fear of slave uprisings—that just before the Civil War the Maryland Legislative Committee on the Colored Population recommended not only against emancipation, but for reenslavement of all free Negroes. There were then almost 25,000 slaves on the Eastern Shore. More than half a century earlier, Judge John Beale Bordley had tried to show that it made more economic sense for a landowner to hire seasonal workers to do the field work than to permanently keep a greater number of slaves and their families. It does not appear that Bordley's arguments caused any masters to free their slaves. Bordley himself, other shore planters must have observed, kept a number of slaves on his own Wye Island plantation. Eastern Shore racial sentiment was probably best expressed in this resolution adopted by lower Shore Dorchester County in 1860: "Maryland can never be the paradise of free Negroism. . . if involuntary Negro servitude cannot exist, we must have exclusively white labor. . . if in the providence of God, this country was intended as a home for the exclusive occupation of the white man, there should be no dark spots upon it. . . it should be white all over."

There are still natives around who proudly remind a visitor that in 1860 Abraham Lincoln received not a single vote in Queen Anne's County. This perspective about the Emancipation Proclamation has always been reinforced on the Eastern Shore. During his research for *Rivers of the Eastern Shore* (Cambridge, Maryland: Tidewater Publishers, 1944), Hulbert Footner had occasion to visit Tilghman Island. He was struck by the politeness in the Land of Pleasant Living, by the way "the old Negroes touch their caps and murmur, 'Gemmen!'"

A local newspaper article of about the same period tried to attract white visitors this way: "Tilghman Island is a White Man's Paradise. . . . One of those places where you are heartily welcomed and splendidly entertained.

Colored people are very scarce, except a few who work in the oyster houses during the shucking season."

Almost a hundred years before Footner visited Tilghman Island, Frederick Douglass explained why, even a century later, black people would be so polite to whites:

The man, unaccustomed to slaveholding, would be astonished to observe how many *floggable* offenses there are in the slaveholder's catalogue of crimes; and how easy it is to commit any one of them, even when the slave least intends it. A slaveholder, bent on finding fault, will hatch up a dozen a day, if he chooses to do so, and each one of these shall be of a punishable description. A mere look, word, or motion, a mistake, accident, or want of power, are all matters for which a slave may be whipped at any time. Does a slave look dissatisfied with his condition? It is said that he has the devil in him, and it must be whipped out. Does he answer *loudly*, when spoken to by his master, with an air of self-consciousness? Then, must he be taken down a button-hole lower, by the lash, well laid on. Does he forget, and omit to pull off his hat, when approaching a white person? Then, he must, or may be, whipped for his bad manners.

The Eastern Shore has generally exhibited a certain schizophrenia about the fact that a lot of black people live there. In Queen Anne's and Talbot counties, for example, a fourth of the population is black, most of them descendants of slaves. One approach has been simply to ignore them. In 1937 the *Easton Star–Democrat* ran an editorial promoting the near racial purity of the Eastern Shore as "90 percent Anglo Saxon." On behalf of the 1933 Kent Island homecoming day this editorial appeared in the *Queen Anne's Record–Observer*: "No country, no people love their home more or their friends with truer affection than the people of Kent Island. The fertile lands of this island have never felt the tread of foreign feet. The ancient blood of the Anglo–Saxon, first brought to these shores by Claiborne. . . still flows after three hundred years almost as pure as it did then."

The other approach about blacks has been to emphasize the negative. Before World War II the *Record–Observer* rarely observed that white people committed crimes. "Officers raid Negro joint. . . [which] has become an offence to the white residents living near the spot." "A Negro fight occurred Monday evening in the vicinity of Queenstown." "Nab Negroes and Locate Lost China." About a crap game: "Bone rolling brings strife to Pondtown. . . when a. . . half dozen young bucks worked up a first class battle royal." "John Adams, Negro, disorderly conduct." Another article spoke of a "blackamoor." In the mid-1930s, when blacks opened their own newspaper in Cambridge, the *Queen Anne's Record–Observer* wrote: ". . . The news content included three and a half columns on the Italo–Ethiopian war, pictures and stories of Noble Sissle and Joe 'Brown Bomber' Louis, also two pictures of local Negroes with the prefix of 'Mr.' carried in the captions."

One might fairly conclude that this is evidence only of the times, when blacks everywhere were treated as second class citizens, when, for example, the Queen Anne's County Garden Club gave prizes to the black residents of Centreville who beautified their yards. On the Eastern Shore, however, racial feelings ran much deeper. In 1933, the same year that a mob of five hundred whites tried to get at a black prisoner in the Salisbury jail, the *Record–Observer* had this to say about a lynching of a black man in a nearby town: ". . . There can be no tolerance for mob law under any circumstances, but until the conditions that help breed them are removed [delays in getting trials and securing convictions, it said], there cannot be serious condemnation."

School integration, therefore, came slowly to a place like Queen Anne's County. Even before integration, when the county was consolidating its many scattered crossroads schoolhouses, there was resistance merely to fixing up the black schools, most of them only shacks. "The schools for the niggers are good enough," a now deceased county commissioner once told former school superintendent Dr. Harry Rhodes, "and we don't need to spend any more money on them." Many white bus drivers came to Rhodes and said, "We're not gonna drive any buses with niggers on 'em." Rhodes simply told them, "Fine, all you have to do is drop your contract," but no bus driver accepted his offer.

In the 1960s the racial pot boiled over, in Cambridge, where four thousand blacks lived in one seedy section of town and nine thousand whites in another, separated by Race Street. Throughout the spring of 1963 blacks in Cambridge agitated against segregated public facilities. There was a large protest march and armed clashes with white segregationists. The National

Guard was called in to restore order. Race relations continued to simmer until the summer of 1967, when H. Rap Brown rolled into town and agitated a cheering rally of about four hundred young blacks to "get your guns" and put the ghetto to the torch. That night, as Brown slipped safely out of town, gunfire erupted and a dozen black stores and the black elementary school went up in flames. It took seven hundred National Guardsmen and over a hundred state troopers and local police finally to contain the riot.

The all-black schools are gone, and Cambridge seems quiet, but the attitudes that time has not erased still prevail on the Shore. Arthur Bryan relishes boyhood memories of how his family's black servants tipped their hats and called him "Mister Arthur." The last descendants of the Bryan slaves left the Wye about ten years ago. One day some of them drove down from New York to visit Arthur Bryan, "in fancy Cadillacs," he recalls. "They put their feet up on my nice antique chairs. I told 'em to get the hell into the kitchen." Bryan thrust out his jaw. "That's how I treat them kind." A few summers ago at a county picnic, a man stepped to the microphone and said: "Aren't we having a good time? And just think, there aren't any niggers here, either!" A schoolteacher from Washington applied for a job at one of the public schools in Talbot. At the interview she was first asked, "How do you get along with niggers?"

Cordova lies inland from the bay, along a railroad spur that once made it a cannery town in the days when the farms around it were solid with vegetables. Since it is not a waterfront village, Cordova has not attracted the well-to-do. It has a small general store and a volunteer-operated fire truck, and little else. Abandoned cars and pieces of scrap metal are in some of the yards. The yards back up to a few dozen frame houses, a number of decrepit shacks, and many trailers.

Trailers. Nothing quite like the sight of trailers is so revolting to those who do not live in them. "Ugh! They're hideous," said a woman of wealth who lives in a handsome home on the water. "They burn like a matchbox, depreciate like hell, and the wind can roll them over," I was told by an official of the Maryland Historical Trust in Annapolis; he was worried that the Eastern Shore, particularly the poorer lower Shore, was being overrun by mobile homes. Of course, many frame farmhouses burn like matchboxes, too, but old farmhouses are part of the Eastern Shore, people who move there say. They are "quaint," "rustic," and "pretty to look at." "But *trailers*, well, you *know* the kind of people who live in them." Admittedly, it is difficult to find anything quaint, rustic, or pretty about a trailer, and some of the shoddy trailer subdivisions in the woods around Easton are little better than slums. But to laboring people, to poor people, to many people on the Shore who cannot afford to buy a house—in no small way because the cost of land has been bid out of reach by the well-to-do newcomers—a house trailer is about the only possible shelter.

That was the circumstance of Peggy and Robert Murray, when they tried to find a place to live in Cordova. The best they could do was to rent a shack that crawled with rats. And that got Lillian Stanford's back up.

Lillian is Peggy's mother. She has lived in Cordova all her life, but, like most black people, has had to live on the outskirts; at least she did until she finally saved enough to buy a lot and build a prefabricated home on it. As soon as she moved in, Lillian's white neighbor erected a nine-foot-high fence between them. Lillian helped Peggy and Robert buy a used trailer, and then she went to Charles Shortall, a local farmer who owns land on the other side of Lillian's lot, and got his permission to move the trailer there. Unfortunately, none of them knew about the Talbot ordinance that requires the filing for a permit, the posting of a notice, and the waiting of ten days before a trailer can be placed on the land. But Joe Secrist, an electrician who lived about a quarter of a mile across a field from the Murray's trailer, knew of the ordinance. Secrist and a group of other white citizens of Cordova immediately raised hell with the county to get the trailer pulled out, so Bill Fleming, the Talbot County zoning coordinator, called a public meeting to air the dispute.

Fleming's office is in the basement of the county library in Easton. Crammed in it the night of the meeting were Joe Secrist and about a dozen other Cordova residents, including Charles Shortall, Lillian Stanford, and Robert and Peggy Murray. Joe Secrist stood up and said that the Murray's trailer was an eyesore, that it was just a few hundred feet behind his house, that it would diminish property values in Cordova, and that the trailer was not lawfully on Shortall's land because a permit had not been first requested and the necessary sign posted for ten days. Secrist had done his homework on the ordinance—he spoke confidently and, when he sat down, he gave a triumphant smile. A number of others spoke up: "That's right, that trailer *is* ugly, an eyesore." "Our property values are certain to go down." Charles Shortall has a kindly face, but he was scowling when he interjected, "That's *my* land the trailer is on—I ought to be able to put anything on it I want!"

Another man got up, squeezed his porky hands together and, glancing back at Joe Secrist for support, turned to Bill Fleming. "That trailer sure would be an eyesore," he stammered. He was uncomfortable speaking before an audience, even one this small. With his thumb he tried to tuck his shirt around an ample waist. "Besides," he said, "I had to get a permit for my trailer—so should they."

The Murrays never spoke. They sat close together, resigned and fearful. Mostly they looked at the floor.

But Lillian Stanford had heard all she wanted to hear, and when she stood up and turned on her antagonists, her eyes were blazing. "You should be *ashamed* of yourselves," she stormed. "I have lived in Cordova all my life and raised Peggy there, and I never expected my neighbors to treat my family, like this. You should see the shack this young couple is living in. . . the squalor. . . rats running through the house! They are planning to have the trailer painted and make it look real nice."

"No one here is trying to get anyone mad," a younger woman quickly said, "And we're all close friends and neighbors, but procedures are procedures, and it isn't fair for us to have to get permits and them not to."

When the meeting broke up, Lillian Stanford, barely containing her fury, led her daughter and son-in-law past her neighbors and out of the office.

The heavyset man who had complained about having to get a permit for his trailer sat in an anteroom with Joe Secrist and giggled about the successful way the meeting had gone. Howard Walker walked into the room, planted his legs apart, and glared at the man. Howard Walker is the plumbing inspector for the county. He is not big, but his glare, when he is aroused, could melt iron.

"How come you're opposed to those people getting a permit?" Walker demanded.

"Well, they should get a permit like everyone else," the man stammered.

"Listen," said Walker, "You got a short memory. You didn't get a permit when you connected that septic tank to your trailer, and it was only because I was willing to bend the county regulations that you were allowed to stay in it." A loose smile fluttered across the man's face and vanished. Walker continued to stare down at him.

The Murrays got their permit.

———— ᢕᢆᢐ ————

Among many people on the Eastern Shore, there is a fear of those who seem different. Outside the Tidewater Inn in Easton one afternoon a group of Japanese businessmen gathered about their luggage, while waiting for their transportation back to Washington, and snapped pictures of the colonial inn that was built in the 1950s. A real estate man, who sells waterfront estates to the wealthy, surveyed the foreign visitors with open contempt. "What are those *Japs* doing here?" he hissed. Slowly with elaborate disgust, under his breath, he approximated the Japanese form of greeting: "Oooo-KODeeeeDESSSSS-*KAH.*" Then he turned on his heel and walked away.

On the upper Eastern Shore, as elsewhere in this country, people resent growth because of the increased population it would bring. But they also fear whom these newcomers might be. A native woman: "Rouse's village on Wye Island would mean a sewer system. . . a dual-lane highway. . . all those kids." And then, to my bewilderment, she added, "They'd all be Bethesda Jews—or Italians. They're the ones with money who can afford those lots." Over the phone, I inquired of another native whether he had certain information about Wye Island. These were his first words: "You're not one of them goddamn Jew reporters, are you? I had one of those goddamn smelly Jews in here the other day, and I still don't have the stink out of the house!"

The distinction between insiders and outsiders is often a murky one. Natives can become the outsiders, even when race is not the issue.

The Bridge Restaurant, for example. It sits along the narrows on Tilghman Island, at one end of the drawbridge. For years it was the gathering place for the island's watermen, who would slog in for beer after unloading their skipjacks of oysters. (In the War of 1812 the British kept a barracks on Tilghman Island and used its sheltered harbor as a roadstead from which they sent raiding parties across the bay.) Oystering is a cold, grueling, dangerous business, and when the men come off the boats, they are ready to unwind. But when Francis Cole, a native of Easton, bought the restaurant, he remodeled it to attract tourists who enjoy sitting down to an excellent cream-of-crab soup and, in warm comfort, watching the oystermen bring in their boats. So Cole threw all the watermen out—told them they were no longer welcome, period. Cole said that the former place "was filthy, and so was the language. . . no woman would dare enter," and it was common to see drunk watermen dragged out like sodden grain sacks. "I don't put up with that sort of thing here," said Cole, who is big enough to serve as his own bouncer. "I'm not running a beer joint—it's a restaurant." Cole refers to himself as "the Toots Shor of the Eastern Shore." Cole agrees that the watermen ought to have some place they can go "with their dirty boots on" after a day on the water, "but it should be down the back streets in a shack," he says. "That's the only thing they're used to."

When Francis Cole ran (unsuccessfully) for Talbot County commissioner, he got few, if any, votes from the Tilghman watermen.

Many of the farmers and watermen resent the rich people from New York, Philadelphia, and Washington, who now command the best waterfront, who travel in Mercedes, who send their children to private schools in the winter and to Europe in the summer. Not a few of the wealthy are content to enjoy the natives only from the comfortable distance of romanticized books about the Shore. Most of the whites, rich and poor, native and newcomer, hope to keep the blacks out of sight in the woods and in their own little settlements. Insiders can be treated as outsiders, and outsiders try to become insiders.

But to most who live on the upper Shore, native born or recently arrived, there is one common enemy, above all others, who threatens the Land of Pleasant Living, who almost single-handedly kindles all the stomach-knotting fears about change. You will not find this ogre mentioned in history books about the Shore, and you will search in vain for it in the *Encyclopaedia Britannica*, but you will hear of it from those who live on the Eastern Shore. Plague of the present and ill omen of the future: *the chickennecker.*

———————— ⤫ ————————

Chickenneckers are, primarily, migratory creatures. Most nest in Baltimore, a city which few self-respecting residents of the Eastern Shore consider fit habitat for anything living, even though many people on the Shore came originally from Baltimore. Before commencing their summer migration to the Eastern Shore, the chickenneckers prepare for the voyage: they acquire a ball of twine, a bushel basket, a long-handled net, and a number of fixed or collapsible wire cages, called "pots." From the grocery store they purchase a package of chicken necks. The more advanced of the species may trailer a boat. Thus armed, the chickenneckers stream across the bay bridge and spread out along the shore to seek their prey: *Callinectes sapidus*, the Chesapeake blue crab. Because virtually all of the waterfront of the upper Shore is privately owned, the chickenneckers tend to congregate around public landings and on small bridges.

Mr. Kaiser, a meat cutter from Baltimore, prefers crabbing from the Wye Island bridge, as do hundreds of other chickenneckers on the weekends, so he usually drives there on Monday, his day off, when he and his nephew can enjoy relative solitude. The variety of crabbing devices that Mr. Kaiser suspends from the rail suggests that he is a man who enjoys experimentation. To the end of a piece of twine he ties a baited, galvanized wire box, which opens flat when it settles to the bottom of the river. As Mr. Kaiser brings it up, the sides snap together, imprisoning a feeding crab, or, more often, nothing more edible than the water of the Wye Narrows. At the end of another twine is a pyramid-shaped crab pot. A third is tied to a steel hoop across which is stretched a fine mesh—a drop net containing a chicken neck.

But for all his gadgetry, Mr. Kaiser doesn't eschew fundamentals. Most of the lines that he affixes to the bridge rail are connected only to chicken necks. When the twine goes taut, signaling a feeding crab, Mr. Kaiser slowly pulls it hand over hand to the surface and attempts to scoop up the crab with a long handled net. Catching crabs in this manner—the standard method of chickennecking—requires quick reflexes, because even the hungriest crab has a sixth sense about the intention of chickenneckers. It can disengage from feasting and swim like a rocket away from the swooping net.

The chickenneckers take home their baskets of crabs, pop them in a steamer, and sprinkle them with a hot spice. When the crabs are cooked to a bright red-orange, they are piled on newspapers and cracked apart with wooden mallets. The meat is succulent. Ice-cold beer chases the spiced thirst. For thousands, catching and eating bay crabs is the source of happy times.

But to people on the Eastern Shore, particularly those who live or work on the water, the chickenneckers represent the ultimate outsiders and everything fearsome about growth. It is common knowledge there that chickenneckers are black, that chickenneckers come from Baltimore, that they leave their trash for the landowners to pick up, that they are rude and noisy, and that they turn their radios up so loud that at night you can hear their parties a mile away. Chickenneckers love crowds. Chickenneckers make crowds. On the landings. Along the roads. On the water. The chickenneckers are all over.

On summer weekends the Wye Island bridge is almost impassable because of people gathered there to catch crabs. Many camp along the road next to Arthur Houghton's fences on Friday night and get up at 3:00 a.m. on Saturday in order to find room along the rail to lay their crab lines. By noon the bridge is aswarm with crab lines, aluminum folding chairs, beer coolers, transistor radios, and people. Those with boats lay crab pots all over the river, each pot identifiable at the surface by a bobbing Clorox bottle. Vehicles jam the road on both sides of the bridge. At the public landings, particularly, such as Wye Landing across from the eastern end of Wye Island, the same scene is repeated, only worse: crowds of crabbers, boats, buoys, boat trailers, campers, and cars cramming the roads and the rivers.

Ray Warner, eighty-seven years old and still working the water, puts it this way: "Saturdays you can't even crab 'cause of the chickenneckers. It's turrible. . . the river's nothing but boats. They even crab nights. . . drag nets around the shore. It's turrible. There's just no peace now. They're from Baltimore City and Philadelphia. I wish you'd go down there at the landing some night. I can't understand it. We got too many people coming here. They're taking away our living. They got more crab lines than we got. They wouldn't like for our people to go up there to Baltimore and take away *their* jobs. It wouldn't be so bad if they just came down for a bushel or so, but they got buoys all over. They catch a boat load, almost. We got a man here Friday night came in with four or five bushels. He went out the next morning and stayed out there until about 2:00 p.m. and caught another boat load. If they just caught enough to eat, it'd be one thing, but they got *commercial* licenses."

Sam Whitby says that it is getting so he can hardly get around Wye Island in a boat. "In just another month the chickenneckers from outside the Shore crowd down so that the people who live down here can't crab. The water used to be so clear and pretty you could see the minnows. My father used to hunt terrapins and could see 'em down in four feet of water."

This is all quite true. But something else is also true. A lot of chickenneckers are not from Washington or Baltimore or Annapolis. A lot of chickenneckers are from the Eastern Shore, from Queenstown and from Centreville, and Easton and Cordova. A lot of chickenneckers are white. Some are lucky to have friends who still live on the water and invite them down to crab from their piers. People who live on the waterfront and catch crabs with chicken necks from their private piers do not consider themselves chickenneckers. But if you do not live on the water, as many natives no longer can afford to do, you have to fight for room at the overcrowded landings with the chickenneckers from Baltimore.

When Talbot County proposed expanding one of its small landings just off the Miles River, emotions ran high:

. . . We are opening the doors to people who don't pay taxes in this county and therefore don't have to care how they treat our facilities. . . . We aren't getting the [state] money without a price and the cost is the type of life we either grew up with or moved here to find.

As a poor Talbot Countian, I am in favor of the proposed project. There are too few public landings in Talbot County to launch my boat.

. . . No one wants to buy property that is in danger of being overrun by out-of-town people. . . .

By having access to the public landings I have been able to enjoy fishing and boating on our waters. . . . Without these landings, how can we get to waters if we don't own property on them?. . . From what I've seen, our visitors rate with our own residents when it comes to littering and abuse.

. . . In a few years. . . the landing becomes a public park which attracts undesirables . . . .

When a new resident argued that the hearing had been unfairly packed with watermen from another village, one of the Talbot County commissioners told him: "You and others buy waterfront property, move in here, and then try to tell a lot of poor people they can't get on the water. This

145 ᧕

landing is being built for the people of Talbot County, and I won't accept any arguments from foreigners."

There is no dispute that the upper Shore's public landings are a crowded mess on summer weekends—too many people trying to occupy too few small landings that were built for an earlier century. The counties have virtually no park superintendents, and the police (usually state troopers, because the counties employ few of their own) only appear when the crowds get overly rowdy. The trash barrels quickly overflow.

But trash is not just a chickennecker by-product. Every morning out over the oyster beds—after the skipjack cooks have fried bacon and eggs for the crews—paper bags, full of egg and milk cartons and other supermarket trash, are slipped over the sterns to float away, sink to the bottom, or wash up on the shore. So, too, from the patent-tong boats and the hand-tongers. Banana peels, orange rinds, paper cups, candy wrappers. Over the side. And the same with pleasure boats and yachts. Over the side. And the cars and pickups. Out the window. Russell Train, the EPA administrator, spends more weekends than he would prefer picking up trash from sightseers along his farm road.

Outsiders, that is what they are, whether they come to catch crabs, explore the rivers, or just sightsee in the villages. "Damned outsiders! We could use fewer of them!" One homeowner in Saint Michaels has found them urinating in the bushes right under his bedroom window. They have also used his lawn furniture without asking, and once, when he was gone, two strangers walked into his house and used his telephone. Not many years ago farmers allowed hunters to walk their fields in the fall and hunt quail. Now No Trespassing signs are everywhere, and the *Easton Star–Democrat* has a lengthy section listing all farms where trespassing with "dog and gun" is forbidden.

Since 1683 a ferry has been crossing the Tred Avon River between Oxford and Bellevue every day all year long. It is the oldest continuous ferry service in the United States, and about the last ferry on the bay. The owner of the ferry recently proposed reestablishing service across Eastern Bay between Claiborne and Romancoke on Kent Island, a distance of about ten miles. This would not only bring back the leisurely pace of the Eastern Shore's past—an era that everyone there strains to recapture—but it would more than halve the driving time to the bay bridge for those in Talbot who must drive far around Eastern Bay.

Many people, however, who bought a lot in Dave Nichols's Romancoke-on-the-Bay do not want a ferry landing there. They say that they moved to Kent Island to relax and get away from city pressures. A ferry landing, they fear, would bring in congestion and more cars and make Romancoke a tourist attraction. It would "disrupt the community."

When Erma Day and her husband moved to Romancoke in 1967, the defunct ferry pier and landing was a magnet, drawing chickenneckers to crab, fish, or picnic. Erma Day remembers, "Riffraff, a very, very low class of people—from places like Glen Burnie and Pasadena [towns between Annapolis and Baltimore]." They threw their garbage everywhere, "relieved themselves in the open," she recalls, "and had sex orgies on our lawn." Shooting drugs, some of them were, she said. "A real nightmare." Queen Anne's County recently took over the landing, fenced it, put in toilets, garbage cans, picnic tables, and an attendant who closes the place at 8:00 p.m. County residents are admitted free, but outsiders have to pay a dollar, which Erma Day says tends to discourage the rougher crowd.

Erma Day is not against the idea of a ferry, "But it shouldn't be in the middle of a community," she says. "It should be up at Kent Narrows, where everything is strictly commercial." A new ferry to Romancoke—the first honest-to-God ferry since the coming of the Chesapeake Bay Bridge (over which Erma Day's husband commutes each day an hour and a half to his job in suburban Washington)—would only bring back the nightmares.

"We've got more than $125,000 invested here and don't want to see that depreciated," she said. "We have no intention of letting a ferry come in here and ruin what we have been trying to achieve here: peace of mind. How could you control the kind of people who would come here?"

———— ✧ ————

The meeting of the Bennett Point–Piney Neck Citizens Association had been underway in the Grasonville school auditorium for about half an hour, when Mr. and Mrs. Eugene du Pont III walked in and sat down behind Brad Smith. They arrived in the manner of celebrated guests at a large dinner party: fashionably late and pleasurably confident.

Frank Hardy had finished his doom-laden statement about a state park on Wye Island. Carl Blakely raised the question of how they might put alternatives "before the medium to counter somewhat, shall we say, the pee-ahr put out by Rouse." Jerome Gebhardt explained how his search committee had found that "you don't get good pee-ahr for nothing. . . and we don't have that kind of money." Blakely added that he and Gebhardt had wheedled one fellow down to seven thousand dollars. Blakely pleaded for donations, but there was little response.

Eugene du Pont decided it was time to hold a plebiscite. He rose and buttoned his blue blazer.

"How many of you have seen the Rouse display?" he demanded. A few cautious arms were raised.

Eugene du Pont clasped his hands behind his back. "How many of you are *for* it?" Frank Hardy and Bill Chaires raised their arms.

"And," asked Eugene du Pont, smiling, "how many of you are *against* it?" The room became a forest of arms and hands. Eugene tweaked the knot of his regimental tie and sat down.

Frank Hardy's patience was wearing thin. "I think it's time to look at the options for Wye Island," he said. "I would be delighted to see Wye Island stay as it is in open space, but we have made a decision to dispose of Wye Island. And this decision," as Frank Hardy had been saying all year since the Rouse option was first announced, "is *irrevocable*."

Frank Hardy folded his arms across his chest. "We can go the five-acre route, but I don't think the people want it. You could easily take the island and sell it off in small farms; some might decide to keep theirs, and some might decide to chop up theirs. There's another option." Hardy paused. "Sale to a *foreign power*. . . there have been some indications of interest." The audience began to murmur. No one in the room needed any further elaboration on which foreign power that might be. The Russians' retreat at Pioneer Point is but twelve miles due north of Wye Island, and Hardy had once rented the Duck House to the USSR.

Hardy continued in his heavy baritone, "You could drop back to two acres and go up to 1,200 units, or finally down to one-acre zoning, and you're up to 2,000 density." He could not resist repeating the ultimate threat: "A state park can still come in. . . and then you've got 5,000 people and 5,000 outboards!"

As he sat down, Frank Hardy said: "If nothing else ever comes of the Rouse thing, at least I hope the plan has shown certain people in Centreville what tertiary treatment is."

Norman Jubb, of the local Watermen's Protective Association, got to his feet. Our organization is flatly opposed to the Rouse plan, he said. "Twenty-five areas in the bay have been closed since the fall because of pollution. The oysters in the Wye are some of the finest in the upper bay. . . . We calculate that the five-acre zone would provide less than half the population of the Rouse plan."

A man sitting in the back row asked Jubb what were the main sources of pollution on the Wye. "The Wye doesn't have marinas on it, no villages on it. We think it's coming from the streams that dump into it," Jubb answered.

"I've heard that fecal coliform from geese is the greatest source of pollution in the Wye," said the man. Jubb swung his massive body around to face him: "If that's true, how come the Wye stays polluted all year?"

Frank Hardy interjected: "The geese stay there a lot longer than six months. How long do you think it takes that stuff to disintegrate in the water?" Sam Whitby once said, "The worst pollution is from the geese in Wye River."

Mrs. du Pont stood up. "Just who are the scientists who say that the Rouse development will not add more pollution to the bay?" she demanded.

Hardy told her that the Rouse Company had a foot-high stack of scientific reports which would be available at the planning and zoning office in Centreville.

Mrs. du Pont then asked, "What is the purpose of a commercial inn that will draw people from Baltimore, Washington, Philadelphia, and Annapolis? It was said at one of the meetings that the Wye Island Association membership would be opened to people other than the residents. What is to guarantee that the association can't decide to change all the covenants?"

Hardy answered that the covenants would probably be limited to, and unchangeable within, a specified time.

Mrs. du Pont said, somewhat bitingly, "*Thank you.* I've just been trying to get these questions answered for some time." She sat down to vigorous applause, particularly from the women.

When the meeting broke up, I asked Mrs. du Pont whether she was opposed to Rouse's plan for Wye Island. She did not answer directly. "Well, as you can see, we are just newcomers here and only trying to help out."

I asked her about the comment made by the man about goose pollution.

"*That* fellow I don't believe is even *from* Queen Anne's County," said Mrs. du Pont who is also not from Queen Anne's County, although she and her husband share a nice view of Wye Island from across the Wye East River in Talbot County. "I'm convinced," she continued, "that he was one of *their* pee-ahr people," vaguely indicating in the direction of Frank Hardy who was leaving by the back door.

"How can you tell?"

Mrs. du Pont informed me. "You can just tell by *looking* at them, how they *dress*. The kind of *shoes* they wear." She smiled, drifted over to her husband, took his arm, and the Eugene du Ponts swept off into the night.

Frank Hardy was on the back steps leading to the parking area. Hardy had often described the people on the Shore who he detests the most: the people who had just arrived and bought their place on the water, and who now wanted to close off the Shore to everyone else. Hardy referred to it as his lifeboat theory. He pointed through the doorway, at the dispersing members of the Bennett Point–Piney Neck Citizens Association. "Those are the people in the lifeboat I was telling you about," he said.

Opinions on the Eastern Shore about the brothers Hardy are, if nothing else, spirited:

They're nothing but a bunch of developers.

I'll make a guess that Frank Hardy goes bust. They've been very lucky. It was because of the money they inherited from their mother.

I remember when Frank first came down to the Shore; he used to work in the fields just like the hired hands.

The Hardy brothers have more nerve and gall than anyone I know. I think they lay around awake at nights thinking how they can get ahead of you.

The county would be a whole lot better off if they'd never seen a Hardy.

Most of the people who don't like the Hardys are just plain jealous of them for making so much money.

Frank Mannen Hardy has piercing dark eyes, as does his brother Bill, and his father William A. Hardy, Sr.—the dark eyes of the Welsh. His father has traced the family back eleven generations to Anthony Hardy, who was born in 1605 in Pembroke, Wales. Anthony's son, John, sailed to North Carolina in 1695 and bought 350 acres. "The most exciting Hardy," says William Sr., "was my great-great-grandfather's brother, Charles Hardy—one of the original trustees of what is now Emory University in Atlanta." William Sr. took a degree in chemical engineering at the University of Washington, once taught at the University of British Columbia, and retired from the navy in 1932 as shop superintendent of the machinery division of the Brooklyn Navy Yard. Shortly before Pearl Harbor, Frank's father was recalled to the navy, and assigned to New London, Connecticut, as senior inspection officer for the supervisor of shipbuilding of the Electric Boat Company. "I took *seventy* subs out for their first dives," William Hardy, Sr., says proudly, "which is somewhat of a record, if you've ever read of submarine trials."

As a boy, Frank snatched rides in the crash boats, and ever since has been in love with boats and the sea. Then came Choate, Harvard (economics), liberally interspersed with sailing, and two years as an engineering officer on a destroyer. For a few desultory weeks Frank endured the executive training program at Marsh & McClennan—his maternal grandfather, Frank Armstrong Mannen, was an early partner of the firm—but Frank's heart was not in selling insurance, and he could not stand New York City. "Commuting four hours a day [from Greenwich] didn't appeal to me," Frank Hardy said. "I went into a bar and got bombed, and came down here"—to the Eastern Shore.

Frank and his brother Bill, who had preceded him to the Shore, began buying farms—Mainbrace, Wye River farms, Bennett Point, My Lord's Gift.

151 ☙

Frank found it hard to stay away from yachting—to Bermuda in the summer, then the Halifax races and Long Island Sound, on sleek racing boats with names like the *Viking* and the *Magic Carpet*. But Kate Hardy, the Easton girl Frank had married, took a dim view of his prolonged voyages, so Frank decided it was time he began making money. He and Bill acquired some ranches in Florida and about 4,000 acres of Arkansas rice land. Bill Hardy divorced and moved to Florida to oversee the properties there, but Frank stayed on the Wye River at Belle Point Farm.

The Hardys had their eyes on Wye Island, and when Jacqueline Stewart died in 1964, Frank and Bill submitted the highest bid and acquired the Stewart interest in the island for about $675 an acre. They planted the Wye Island pastures in corn, and with the corn came more and more geese, and, ultimately, Rockwell International for the gunning lease.

The Hardys discovered that raising livestock and cultivating crops was risky and required a pile of money: tractors, plows, combines, seed drills, fertilizers, pesticides, feed, wrenches, hammers, fencing staples, wire, shovels—everything from $30,000 machinery to bolts, nuts, nails—hundreds of different things that wear out, get lost, rust, or break down for no explainable reason during the moment of harvest, when a lost day can mean a frost and a huge financial loss.

The Hardys wanted more money, not only to keep their present farms going, but to expand, to buy more land. They wanted all of Wye Island. Frank Hardy knew that trying to talk Arthur Bryan out of his 150 acres at Bordley Point was a waste of time, but he knew there was a chance to get the rest of the island, Wye Hall Farm. That meant convincing Ruth Brown Dorlon to sell off the estate that she and her late husband had held since 1944. But Ruth Dorlon was widowed in 1966, her three children were grown, and Wye Hall Farm and its empty mansion had become a burden. Her children argued against selling to the Hardys, but $50,000 was $50,000, and the Hardy brothers were willing to pay her that much for the option to buy Wye Hall Farm (at a price, if exercised, of $2 million). Ruth Dorlon signed, and after the unexercised option expired, she agreed on another. This option was to cost the Hardys $75,000 over three years in order to buy the same place for $2.5 million. With this second option in hand, Frank Hardy signed the agreement with the Rouse Company on May 1, 1973.

"I've acquired the name of mad-dog developer because I developed sixty-four acres [the Mainbrace farm], then a hundred acres at Governor Grason Manor. I don't think that's fair," said Frank Hardy, as he drove toward his subdivision at Bennett Point.

"We created a need for funds to improve those ranches in Florida so I did a little subdivision. Now I'm doing Governor Grason Two. But I'm not primarily a developer. The main reason I got into the development business is that I like Wye Island and want to avoid what the unscrupulous bastards have done around here." We cruised past the entrance to his Wye River home at Belle Point Farm, past Ray Warner's little roadside house, and past the lane leading into the farm of the ultimate Wye Island holdout, Arthur Bryan.

"We are primarily an agricultural and cattle operation," Hardy said. "We were the first to grow rice in Florida." We neared the end of the county road, the entrance to the Bennett Point subdivision.

"I bought Bennett Point to *protect* Wye Island," Hardy continued, a statement he would repeat twice more that day. Hardy was carving Bennett Point into five-acre lots and, so far, they were not selling well. His theory of protecting Wye Island by buying Bennett Point went something like this: by attempting *unsuccessfully* to sell off Bennett Point in five-acre lots, Hardy would thereby demonstrate to the county commissioners that Wye Island, as similarly zoned and developed, would be a bust, a carved-up, part ghost-subdivision like so many on Kent Island; this would help convince the county to accept cluster zoning—and Rouse's plan for Wye Island.

"If I was to do Wye Island in five-acre chunks, how would it sell?" Hardy asked. He answered his question. "Hell, it *doesn't* sell! Unless you get $5,000 worth of equipment or a mowing service, you'll spend your life mowing grass!"

Frank Hardy is given at times to exaggeration. But mowing grass is not a minor irritant to city people who move to the Shore. Retired business executives who buy their paradise on the water despair at the inordinate amount of time they find themselves spending on a power mower. (Scott Ditch believes most of them love it, imagining themselves as weather-beaten farmers spring plowing on a big John Deere.) To avoid this, most Shore buyers get the state of Maryland to plant their vast lawns in loblolly pines. A lot of subdivided fields of the upper Shore are becoming pine groves in just this manner. The Maryland Department of Natural Resources, once satisfied that the trees will serve a "conservation purpose," provides the seedlings free. The landowner merely pays for the machine operator, who plugs the foot-high pines into the fields like rows of corn.

Frank Hardy was frustrated and angry at circumstances that were affecting him where it counted: his pocket and his pride. A major recession was underway. Money was almost impossible to borrow, except at prohibitive interest rates. The Arab oil embargo had Hardy's potential lot buyers stacked up in mile-long gas lines in the cities, wondering how expensive it would now be to commute to the Eastern Shore. Frank Hardy's land sales business was barely limping along. On top of this, the county's mood had become so skeptical of development that the temporary subdivision moratorium gave signs of becoming a permanent one. Hardy could feel tremors beneath the Rouse project. He had brought to Wye Island a respected and imaginative developer—and a native at that—but no one seemed willing to listen. Frank Hardy's pride was being assailed.

Also, if Rouse fell, $8,850,000 would not find its way to Belle Point Farm.

We drove through the gates into Bennett Point. The inside of the car reverberated with Hardy's baritone anger at the five-acre zoning enveloping most of the land that he and his brother owned. (It was a wrath that would reverse direction in order to defend with equal indignation the five-acre zone, when events later took a different course.)

"I think the five-acre zoning is fragile," he said. "It will be gone before you know it. This type of restrictive zoning isn't to better the environment, which it does not do. All it does is preclude the utilization of the waterfront. The five-acre limit was put in to maintain a low tax base for a few people under the pretext of historic preservation and that sort of thing. And what you now have with it is *price* discrimination. On good waterfront, the cost of five acres means that a minimum investment in building a home there is $150,000." Frank Hardy said that much of the pressure in the 1960s for five-acre zoning along the Chester and Wye rivers had come from people who wanted to hold growth down in order to keep wages low. One influential woman, who owned a large estate, had told Hardy that if the county ever started growing, she would be unable to afford her servants.

Hardy stopped the car. The road down Bennett Point was blocked by a backhoe clawing a deep ditch for a drainage culvert. Hardy's crew was improving and realigning the road for better access to the lots. Rolls of telephone cable lay about the ground. A group of men stood on the road and watched the backhoe operator digging the ditch. Hardy got out and walked over to Bill Chaires, his sales manager. Chaires informed him that the backhoe had inadvertently cut the telephone cable, which the telephone company had recently buried in the wrong place.

Frank Hardy fairly thundered, "This work was supposed to be done January *first*, and we *paid* for it last *fall!* Twenty-nine thousand *bucks,* and *who* takes all the grief when it's not done? I have an airline pilot who bought a place here and couldn't get phone service until they finally jerry-rigged a connection to him."

Frank Hardy stood in the middle of the road, his face darkening. A man from the telephone company quietly got in his car and turned out into the field to bypass the ditch and head back up the road. Hardy stared at the car for a moment in disbelief, and then his voice exploded like a cannon-shot.

"*Wait* a minute! You're driving across my *corn!*" Hardy spun around to face Chaires. "The sonofabitch! *Look* at that!"

Frank Hardy tends to walk in long, purposeful strides, although he is not tall, nor long legged—he is thick shouldered, and narrow waisted—and this is how he marched over to the telephone company car that had stopped in the field of newly seeded corn. The driver rolled down the window. In measured but polite tones, Hardy informed the man about the corn he could not see. He directed the driver around the backhoe in the direction the man was apparently going anyway, and the car roared away.

I had asked if we could see the rest of Bennett Point, so Hardy motioned me and Bill Chaires back into the car. Hardy swung the car out across his cornfield along the path of the telephone company car, back onto the road, and headed toward the point.

Bennett Point is long and narrow like a witch's finger. Eastern Bay to the west and the Wye River and Wye Island to the east seemed quite close. Flat cornfields, plowed and vacant, ran smoothly to the clustered banks of thick, old trees. There was no sign of human habitation, past or present, until Hardy pulled up to a modern house, which squatted in a vast lawn along Eastern Bay. We got out of the car, walked to the bay and looked out across it. It was early May, crisp and clear, and a brisk wind pounded waves against the white rocks below our feet. We could see Parson Island and beyond it the long silhouette of Kent Island. A few small boats cut through the water.

The white stone at our feet was riprap, almost a mile and a half of it. Frank Hardy had emplaced it along the shoreline to keep the rest of Bennett Point from ending up in Eastern Bay.

"All that riprap cost twenty dollars a foot. Now it's up to twenty eight or thirty," Hardy growled. A hundred and fifty thousand dollars' worth of rocks. Hardy turned and walked into the house.

Hardy built the house for speculation on a six acre lot. From the empty living room, we looked again at the bay through sliding glass doors.

"We're asking $160,000 for it, but I'd take $150,000 today!" Frank exclaimed. He asked Bill Chaires, "Any little monsters writing on the walls in here, Bill?" Chaires said no.

Hardy shoved his hands in the pockets of his tan poplin trousers. He was wearing a blue sports coat, a light blue, button-down shirt, a dark blue tie with small red figures on it, and leather moccasins with deck soles. With or without the tie, this is what Frank Hardy normally wears all year around.

Hardy stared out at the bay and said, "Y'know, this place is a real show in the winter. We banned hunting here. The geese need water all the time, and they know right away when it's safe. You open those glass doors and they'll walk right in the house."

Hardy looked about the bare living room and the blue panorama of the bay. "You'd think the successful retired businessman would he in here," he said.

"But these houses are nonsalable!" Frank Hardy thundered, "*Nonsalable!*"

We drove on down the road and soon encountered a young man and woman energetically digging a ditch. It was the airline pilot and his wife who had had so much trouble getting their phone service. They were among Hardy's few lot buyers on Bennett Point. That night they attended the meeting in the Grasonville school to hear pleas to block Rouse's development of Wye Island, which lay just across the river from their new home. Hardy seemed caught in an ever-ensnarling web: with every lot he sold, he bought an antagonist against further development.

Hardy rolled down his window. "No use your doing that by hand. I'll send one of my boys down here with equipment to do it for you." They smiled and waved as Hardy drove off.

We drove past a clump of weeds enclosed by a rail fence. Obscured within lay the grave of Richard Bennett III, of whom Dickson Preston has written in *Wye Oak—The History of a Great Tree*:

> He was undoubtedly Maryland's first multimillionaire. In the first half of the eighteenth century, he was its greatest landowner, its largest shipowner, its most important merchant and financier. . . . Bennett collected land and estates with a passion which seems almost neurotic. . . some internal need seemed to drive him to continue to collect more land, more money, and more economic power.

The road ended in a grove of trees and underbrush. We walked to the edge of the Wye River. Frank Hardy's mood shifted from pessimism to optimism and back to pessimism again. He spoke of the lagging sales on Bennett Point. "We screwed up immensely the first year. Place has no roads. . . everything in crops. The hordes came down here, and we weren't ready for them. We had a terrible spring, and the road wasn't finished until August. We got off to a bad start. We were trying to sell Cadillacs with no brochure. We've sold only eight lots. It's not a question of price. You could not buy better waterfront. Of course, you need some trees, but they grow fast. Ask a man who's owned one. . . Chaires here."

Bill Chaires shook his head and smiled. "Yeh," he said. Bill Chaires looks like a robust, one-time fullback: silvery hair, wide reddish face, and thickly built. He recently retired from the navy as a captain, but he moves with the natural bounce of a born salesman. He is a native. His father sold farm machinery and the hydraulic lift under the Duck House to the Stewarts.

Frank Hardy speculated how he might attract yachts into Bennett Point. Hardy had told me many times, as had others, that the only way to see the Shore was from the water. When they see it that way, they want to buy. Hardy began to describe how he would construct a small aquatic billboard using pictures from his Bennett Point brochure.

"We'll get a couple of pilings and screw-bolt it down so the bastards can't steal it," he said, "and put on one of our best centerfolds and varnish it so

157

it'll last in the rain." Hardy and Chaires looked at one another and smiled contained "eureka!" smiles.

Hardy gazed across the river at Wye Island's Bordley Point. "I looked up developers all along the East Coast, including Charlie Fraser at Hilton Head. We could have done Wye Island in the usual five-acre bit, throw in one million for a crappy golf course, and high sell out the rest as waterfront lots. Add it up. That's twenty-five million megabucks. But I thought Wye Island was such a unique place, that I looked up Rouse. We went around and around for awhile." Hardy described their negotiations over the option. "Rouse had the unique ability to get the best brains in. . . the best."

Frank Hardy's brown eyes have such a piercing quality that they seem to throw shadows over his face as his moods shift. When he glowers, as he now glowered, it is as if a storm has descended and darkened everything below in melancholy shadow. "I think at this point, quite frankly," Hardy said somberly, "that Rouse's chances with Wye Island are less than 50 percent. If it goes into a state park, I'll sell off everything I have. It's a known fact that it would depress values 30 to 40 percent."

Earlier Frank Hardy had burst out in frustration, "I could be planning to put the Taj Mahal on Wye Island, and they would oppose *that!*"

"Why?"

"All the big landowners took a killing in the Depression. The farmers had no reserves, lost their money, made distress sales, and the big boys came rolling in from New York and Philly and bought up all these places, and the natives were evicted. Since then, there's been a deep resentment against the newcomers. They get it from their daddies. Now they see all this land they once owned selling off for fantastic prices.

"It all boils down to: 'I haven't got *mine,*'" he concluded.

———— ✧ ————

The ownership of land is charged with raw emotion. On the Eastern Shore these emotions emerge most distinctly in the conflict between city people, who sell their homes in Washington at high prices and retire to large waterfront estates, and the natives, who then find the cost of land and basic housing soaring out of sight. The resentment runs deep, for it is born of different lifestyles, of those who read only *The New York Times* and those who read the weather. Different perspectives and different motives.

And not just on the Eastern Shore, but in dozens of towns and countrysides elsewhere in the United States.

Talbot and Queen Anne's counties were trying to keep from being overrun, from being changed too fast, and their principal instruments to control the numbers and types of people moving in were subdivision and zoning ordinances. The debate centered on how big the lots should be to "protect open space"—and to keep people out. Talbot County came up with something called the Village District Zone: it would allow small lots in the villages to sop up the population growth, but zone the waterfront in large lots to keep it from creeping there. The reactions to the proposal went like this:

I don't want a lot of out-of-towners coming in anyway, so hold it to two acres so they won't come in.

Listen, I'll gladly give you a portion of my quarter-acre just to get the grass cut.

I say people have got to have a place to live, regardless of how much land they have to have. They've still got to have a place to live. It's getting so people can't go out here and buy a half-acre for a building or a house.

We just left Prince Georges County [suburban Washington], where we had a lot of nice little half-acre and quarter-acre lots, and I don't ever want to see another one.

You are going back to rich people who want these five-acre lots. That is who is running Talbot County today. Millionaires who have moved in here on us, and we are doing what they want to do.

I'd like to see the area remain rural, but the opposition to the village concept is a serious mistake for Neavitt. You've all got children, older people want to come back to Neavitt, and we don't have any land for anyone to come back to. Neavitt is screening these people out moneywise.

159

No one ever zones property based on who owns property.

If you're within the red line [of the village district]. . . and would not like to be, how do you go about [changing] it?

Well, just tell us, and we'll take you right out.

OK, my name is Biro, and I want to be taken out.

Will you indicate it on the map?

I certainly will.

Bozman is a tiny village near Saint Michaels. The waterfront around Bozman and Saint Michaels is some of the most highly prized on the Eastern Shore. The hearing over Bozman's proposed village district was held at night in a meeting room of the church. It was jammed, standing room only, and little of that. People stood on their tiptoes for room, for air, to see who was talking. Most there were older, and dressed in the understated manner—herringbone jackets and gum shoes—that well-to-do people affect in country places. They talked a lot about protecting Talbot's beautiful shoreline and keeping the area from becoming "overrun." They spoke often of the Land of Pleasant Living.

A few younger men leaned against the wall, looking somewhat out of place. Their faces were blistered by the sun, their hands rough and raw from working on the water. When these men heard the older men in crested blazers refer to the Land of Pleasant Living, they laughed, and said shit, not very inaudibly.

One of the watermen's wives grew agitated as she listened to the retirees from Washington and Philadelphia equate small lots with undesirables. She finally exploded: "What about the people who have lived here all their lives?"

The meeting was steamy and chaotic. Energized by the proposal to create more small lots in Bozman—to allow in more people—the debate ricocheted from one resentment to another. One of the men in the audience took this opportunity to complain of the chuckholes that the school bus was leaving on the lane past his waterfront home. The native woman groaned. "*Some* of us have children that *do* have to get to school, you know," she snarled. The man looked at her, bemused, and straightened his sports coat.

When the meeting ended, the woman followed the man to his new sedan parked outside. They argued back and forth, the woman thoroughly sore and the man indifferent and composed, as if he were calmly waving away a moth. Just before he shut the car door and pulled away, the woman said in an exasperated, somewhat desperate voice: "Well, you've already got yours. We want *ours*, too!"

———————— ⌒⌒ ————————

Frank Hardy drove back up the road, around the backhoed ditch, and out the gate of Bennett Point. At an intersecting lane he stopped the car and pointed to a large old house in the middle of a neglected field. The lane led past the field to a public landing on the Wye, where once a locust pole had strung the rope ferry to Wye Island's Drum Point across the river. The chickenneckers now swarm to this landing in the summer. No Parking signs stood along the edge of both roads. Hardy said that he had counted as many as 225 boat trailers parked along the roads up to the landing.

"Backed all the way into that little old lady's house," he said. "Terrified her. I love for my kids to catch crabs, but I've got no love for chickenneckers. A very unpopular breed. In lousy little outboards."

Among the various boats Frank Hardy has moored at his pier on the Wye at Belle Point Farm is a sixty-five-foot leviathan called the *Coastal Queen*. For half a century, before Hardy bought it, it was used as a buy-boat for oysters. A fat smokestack protrudes from its new superstructure. Hardy likes to have friends aboard for a drink on calm summer afternoons. The topsides are five inches thick, and Hardy is unworried about collisions with other pleasure boats on the bay, particularly lousy little outboards. "If I ever run into another boat," he said, "I know who's gonna win. Ain't *no way*!"

Hardy drove on. At another intersection he turned off and came into a scattered collection of houses on the Wye River. "I want to show you something," he said. "Here is where all the static's coming from over Wye Island."

He parked at a small landing on the Wye. The houses, somewhat new, were part of a subdivision called Wye Acres. Within about two hundred yards of shoreline, I counted five piers.

"This is the kind of thing you see all along the shore," Hardy said disgustedly, swinging his arm in a wide arc. "This is what *everyone* wants—their own little dock and their own little boat," said Frank Hardy, who also owns a dock or two and many boats. "It's what we are trying to prevent with Wye Island. These are the Johnny-come-latelys who are doing the most hollering, and for the life of me, most of them have just arrived. And they never quit hollering."

The nearest house was separated from its pier by a neatly cut lawn. I visited there the next day. The owner was the man who had tried vainly to find someone to run the Bennett Point–Piney Neck Citizens Association's PR campaign against Rouse. Jerome Gebhardt.

———— ❧ ————

Jerome Gebhardt is a big, friendly man. Tall and rambling with wide shoulders, big hands and feet, his movements are easy and fluid. Gebhardt looks as if he played football when he was younger, which he did—semipro ball for a team in Ellicott City, near Baltimore, that once drew as many as five thousand people to a game. His face is large and open and as comfortable as an easy chair. He thoroughly enjoys the company of others, and he loves to talk. Gebhardt's friends call him Gabby, and he seems to like the nickname. He talks in a soft voice that rapidly gathers momentum. Jerome Gebhardt is sixty-five years old, and his gray hair, which he combs straight back, is stiff thick.

Jerome Gebhardt was born in Annapolis. Three years later his father died, and Jerome was raised by his mother and older brother. In 1926 the Gebhardts moved to Catonsville on the edge of Baltimore. "It's *Catonsville!*" he insisted. "*Never* Baltimore!" Gebhardt found work as a machinist for a company making machinery that shaped, glazed, and packaged donuts. He and his wife were married at the beginning of the Depression, and for six weeks he was laid off. But his mind kept working. Gebhardt was naturally inquisitive, and he liked to invent and build things. He made his first and only toolbox out of oak. He built a machine that sharpened lawnmowers, and he got into woodworking to earn extra money. He took night courses at the University of Baltimore in mechanical engineering and industrial management. One night he walked six miles to collect a dollar to buy his family's dinner (hot dogs). "Times like that make you appreciate the importance of saving money and being able now to enjoy the good life," Gebhardt said.

As the years passed and the growing Baltimore suburbs gradually closed upon towns like Catonsville, the Gebhardts talked of some day enjoying the "good life." What they had in mind was leaving Catonsville, for the good life was somewhere else. In 1965 they finally found it: a little over an acre on the banks of the Wye River. The Gebhardts bought the lot and three years later built their present home on it, commuting on summer weekends across the bay bridge. In 1972 Jerome Gebhardt retired as superintendent of the mechanical division of DCA Food Industries, and he and his wife pulled out of Catonsville to live on the Shore for good.

When the Gebhardts built their home, they also installed a doglegged pier. In the summer they place a large live-crab encloser on the bottom of the Wye River and tie it to the pier. The Gebhardts love to catch crabs from their pier; those not eaten in the summer are stored in their freezer for the winter. The Wye is ten feet deep off the end of the Gebhardt's pier. Ten feet is considered to be deep water on the Eastern Shore. Deep water means having your own boat within sight of the living room.

The Gebhardts now enjoy the good life. They own a cabin cruiser. They belong to the Kent Island Yacht Club which, Jerome Gebhardt says, has "a nice class of people, not too poor. . . oh, a few rich." Gebhardt has built a collapsible blind which he bolts down to the dogleg of the pier each fall, when the waterfowl season opens. He keeps a portable heater in it. Their home is newly furnished, immaculate, and orderly. An American eagle emblem hangs over the breakfront. Over the mantle of their fireplace Gebhardt has hung a slab of tulip poplar the size of an ironing board, which he has stained and polished to look like cherry. Sliding glass doors lead out to the porch and to the river beyond the grass.

Jerome Gebhardt stretched out in a living room chair, his large frame clothed in worn corduroy trousers and shirt, his feet in leather moccasins. Gebhardt was happy in his new life. "And I have air conditioning, too," he added. "This really *is* the land of pleasant living."

Jerome Gebhardt is full of energy, restless, and always moving. When we walked outside, I was struck by the neatness of the vegetable garden. It was a perfect rectangle, as if laid out by a surveyor, and its grass border was clipped with the exactness and care of the Paris Tuilleries. The soil had the even texture of gardens I always see pictured on seed boxes in nurseries but am never able to duplicate in my own yard. Not a single weed was evident in the Gebhardt garden, not a single growing thing appeared that was not intentionally planted. I marveled at the care he gave his garden and asked if he spent many hours to keep it that way. "No," said Gebhardt, "I can do it each day in a few minutes. That's just me. I like to get things over fast. I've always liked modern things. Y'know, get going," he said, rotating his flat palms, like lawnmower blades. He owns a Gravely power mower that cuts a forty-inch swath. Gebhardt told me that it takes him "only three hours" to mow the grass each week.

"Retirement can be a problem," Gebhardt said, "particularly when you've been a boss or superintendent. You're under your wife's feet all the time, and she isn't going to put up with bossing. But I'm busier now than I ever was when I was working. After reading the paper, I get going on a long list of projects. There's a complete workshop under this house." I asked if I could see it, and he jumped up and led me down the basement stairs.

"My machine does six arrows at a time," he said, as we thumped down the stairs. "I'm known all over the world: 'Oh, you're *Gebhardt*, the multi-fletcher man,' they say. It's not a get-rich business. It's just a small business, but it gives us enough to buy boats and frivolous things like that for our vacations. And when I sell the business, I hope to make enough for the next ten years."

The Gebhardt basement was as immaculate and orderly as the rest of their home, and exactly as big. The joists above were high; Jerome had insisted that the builder construct them that way so he would not have to stoop. A variety of household items, including Gebhardt's folded-up blind and four bushel baskets full of duck decoys, were stacked about with the orderliness of a marine barracks. Along two adjoining walls was a long, L-shaped workbench, over which were ordered Gebhardt's wide array of tools. His oak toolbox sat on the bench. Near it was a small, prism-shaped, metal frame that held six clamplike devices. It was the multi-fletcher which

Gebhardt had designed, and builds and sells, and for which he is renowned in archery circles.

"Archery is the greatest sport in the world. Other than football, maybe. Archery is a family sport, you know. Archers are getting better, and they try to shoot the feathers off each others' arrows. At thirty to forty dollars a dozen retail for good aluminum target arrows, you're not going to throw the arrows away. You get my multi-fletcher and refeather those arrows."

Gebhardt unsnapped one of the clamps from the multi-fletcher frame. He explained the importance of his device in placing—fletching—the feathers on the shaft to make accurate the arrow's flight. When an archer draws the arrow to his cheek and releases it, the arrow bends five different times—sideways three times and twice up and down—before leveling out in flight. If the feathers are not properly fletched on the shaft, the arrow will not go where the best archer in the world wants to put it. Feathers for arrows come from the wings of turkeys. Feathers from the right wing bend one way and feathers from the left bend the other. Gebhardt says that some archers want left wing arrows and some want right wing arrows, but there are three schools of thought—to others it makes no difference. Gebhardt's clamps curve either way, for right- or left-wing feathers. The arrows are laid in grooved slots in the frame. Each feather is placed in a clamp, a bead of glue applied to its shaft, and the clamped feather then fitted in the frame against the arrow shaft to dry.

Jerome Gebhardt is proud of his design. For some time the Franklin Institute for the Blind in Philadelphia bought Gebhardt's multi-fletchers for those blind people who wanted to fletch arrows. For a basement business, Gebhardt has quite an operation. The plates, legs, and clamps of his fletchers are made by one firm, and the other parts are contracted out to individual machinists. Another firm plates the steel clamps and nock receivers and etches the aluminum frame, and still another assembles, packages, and ships the multi-fletchers.

Were it not for James Rouse, Gebhardt would have been content—enjoying his multi-fletcher business, shooting in archery competitions, and watching the geese fly over Wye Island.

Jerome Gebhardt walked to the end of his pier and stared across the river at the island's Bigwood Cove. "I would ask Rouse, 'Why are you putting two hundred boat slips here, right across from my property instead of one hundred slips on this side and a hundred on Dividing Creek?' This area of the river can't take two hundred boats. The water gets so cloudy now in the summer with fifty boats that it is just awful. The more boats you bring in, the more you devaluate property. We've had people come down here and picnic on our dock. I ask 'em what they're doing and they say, 'It's free isn't it?' "

The gentleness went out of Jerome Gebhardt's voice. "Free?" he groaned, echoing the natives, "*Free? All the city people think everything is *free* here in the country!*"

Gebhardt pointed up the river to the substantial home of Herndon Kilby, the developer of nearby Prospect Plantation. "Now what Kilby is doing is ok," said Gebhardt. "He's got a lot of five-acre lots—estates. Hardy's not having any trouble selling five-acre lots on Bennett Point. Five acres will give you in some areas, and one-acre in others, a good use of land." He referred to his acre on the Wye as an estate. "Density is the problem," he continued. "We don't want more population."

He paused for a moment, then he said that the solution for Wye Island and the rest of the Eastern Shore was simple. It was not villages. "Estates. Five-acre estates."

Gebhardt allowed that one-acre subdivisions were not the best type of development for the Shore. But since an acre was what he could afford when he bought, and what he now wanted was as few people as possible to follow his example and move to the Shore, Jerome Gebhardt's present standard was nothing smaller than five-acre lots. He said that his neighbor, who bought four years after Gebhardt, had to pay almost three times more for his acre of ground than the Gebhardts had paid.

At those prices, I asked, who would be able to afford five acres?

"Who cares?" he answered. "*Who cares?* Listen, there's no one who can live on the waterfront now but the wealthy. The Shore should be a place for the retiree." Gebhardt turned and gestured toward the little boat landing between his pier and his neighbor's. "It's like these people down here," he said, referring to the chickenneckers. "They just crowd up, throw trash. You want *those* kind of people around?"

We returned to the house and pulled chairs up to the kitchen table. I asked him what the county should do about its economy to create jobs, particularly for the natives. Gebhardt was not too concerned about the county's economy: "They don't need all that money from development. The workers here have more than enough work to do. We're not inviting in more contractors and outside people to generate outside work. The young people move away because they don't like the kind of work here. But they come back," said Gebhardt.

He grew silent and thought for a moment, then he frowned. "Who says they *should* move back?" I mentioned the natives who had told me how badly they wanted their children to stay on the Shore, or return, and live near them. "Well that's too bad," Gebhardt replied. "Others will move into their place."

During the debates over Talbot and Queen Anne's counties' proposed zoning and subdivision ordinances, many farmers had complained of the larger lot requirements. They wanted to be free to subdivide a few lots when their cash reserves dropped low (most farmers on the Eastern Shore, as elsewhere, heavily mortgage their land, their principal source of wealth, to pay for expensive machinery and each year's higher-priced seed, feed, and fertilizer). They also wanted to be able to carve out a small lot or two for their children, who they hoped could be encouraged to stay on the Shore, on the farm. And some hoped, eventually, to sell their farms for development prices.

For the farmers, Gebhardt had no sympathy. "Somebody's got to draw the line," he said. "Farmers've got to decide whether they want to be wealthy or stay farmers." Frustration edged into Gebhardt's voice and wrinkled his face. Finally he boiled into momentary anger. "I don't give a goddamn what the farmers want, because they're a minority now! Everyone has certain principles about proper esthetics, environment, and land use. What would you consider a delightful way of living in an area that hasn't yet been commercialized? I want it developed *this* way," and Gebhardt spread his long arms to indicate his home, his pier, Wye Acres, his good life. "*This* is the way we like it. I admit, we're selfish. We moved down here to get away from all that."

Across the river, Wye Island caught the last brightness from the setting sun. Jerome Gebhardt, a man who had worked hard to get where he was now, gathered up all his memories and his hopes and expressed them in one plaintive question, "Why can't there be at least *one* county where things remain as they are?"

CHAPTER 7

THE AUCTION

May 28, 1974. Two and a half months had elapsed since Jim Rouse had stood before the Queen Anne's County Planning Commission and urged upon the county his proposal for Wye Island. Two and a half months of silence. On April 19 Doug Godine, Rouse's point man for the project, had written Robin Wood, the planning administrator, to ask that an informal meeting be scheduled with the county officials and staff so that the Rouse staff could present and discuss its detailed studies. For more than a month the letter went unanswered. "We're being terribly coy with each other," Robin Wood admitted. "We floated that letter past the planning commission, and their general reaction was to do nothing, to keep at arms' length." The county officials sensed that Rouse's project was petering out.

They were right. More accurately, the project was collapsing, not so much of its own weight but of combined forces upon the Rouse Company (and thousands of other firms) brought about by a national economy that seemed to be standing on its head. The oil embargo was having its effects.

Inflation was at 13 percent. Two weeks earlier the nation's leading banks had upped their prime lending rate to 11¼ percent, the tenth such increase in the ten weeks since the Wye Island project had been unveiled. The Rouse Company and its subsidiaries, which had been able to borrow immense sums of money to finance their projects, now found the money market almost impenetrable. Because of materials shortages and inflation the building industry was staggering under costs increasing as much as 18 percent within a single year. Stock prices were at near-historic lows and still falling. The president of the United States was on the verge of being impeached. Italy was practically on its back, and the economies of Britain, France, Japan, and the other industrialized nations of the world were tremulous. People wondered if the world were not coming apart at the seams. Confidence was draining out into space. This was illustrated in millions of decisions across the nation, as families and businesses cut back spending what little money they had.

The Rouse Company was in trouble. Not terminal illness by any means, but it was in trouble. The company found itself caught in a suffocating squeeze between increasing opposition to many of its projects, the energy crisis, escalating costs of construction, and a punitive, if not dust-dry, money market. The company's common stock, traded over the counter, dropped from a high of thirty dollars a share in the late winter of 1972 to about four dollars a share by May of 1974.

The economics of the Wye Island project now dominated the discussions among the management and board of directors of the Rouse Company. It would require a huge investment to create the unique environment on Wye Island that Rouse so passionately believed was the Shore's only hope for avoiding sprawl. While costs soared, due to double-digit inflation, the project's anticipated revenues had shrunk with the lowered density of the plan. Jim Rouse still believed that in time he could persuade the skeptics in the county that the project was in their best interests, but many of his fellow directors felt that Rouse was mistaking politeness on the Shore for support.

Although the Rouse Company still had another month before having to exercise the Hardy option or lose it, there was no evidence that the gloomy forecast for the project would not be even worse by then; the commissioners remained adamant about retaining Wye Island's five-acre zoning. The board of directors met and decided to cut the losses and ride out the economic storm that was descending upon the company and the country. The Shelby Farms project near Memphis, Tennessee, which had also run into virulent local opposition, was aborted, and the company's New Communities program was disassembled.

The Rouse Company was pulling out of Wye Island.

Jim Rouse was disheartened, almost depressed. Though he had initially been skeptical about the project, he had become its most energetic advocate. "This is a real sad day for me," he told his board members. He went into his office and drafted a statement to the Queen Anne's County commissioners.

——————⌀⌀——————

Late that morning Scott Ditch and Doug Godine drove across the bay bridge and into Centreville. They parked across from the courthouse and went into the room where Julius Grollman, Leonard Smith, and Jack Ashley were gathering for their weekly meeting. Godine asked for permission to read Jim Rouse's statement: "We are here today to bring you a report which may bring comfort to you, but which is disappointing to us. . . ." He was interrupted by Leonard Smith.

"That makes me very happy," Smith said, grinning.

Godine went on. "We have withdrawn from the prospective development of Wye Island and have advised the owners of the land of our termination of our option agreement to purchase a portion of the island."

The statement was three and a half pages long. In it Rouse reviewed his initial letter to the county of his intentions to produce a place of beauty and charm on Wye Island, and described the many months of ensuing effort to create a plan to do just that. He talked about the growing congestion in the money market, the persistence of the energy issues, and the "widespread economic uncertainties [that] have been adverse to our economic projections for the project. . . . The absence of encouragement or support from local government has made it clear that the zoning process will be extended and, at best, is uncertain. Thus we find it a poor risk to continue and have no choice but to abandon our efforts." Rouse spoke of his desire to create a unique development of

Exceptional high standards... which could induce and compel high quality development in Queen Anne's County and elsewhere throughout the Chesapeake Bay region. . . . We really believed that the carrying forward of Wye Island as proposed in the booklet which we distributed to the residents of the County would be of enormous benefit to the County. . . . We hoped that you might see these possibilities with us. But you have been unwavering in your opposition. We respect the integrity and conviction with which you seek to protect the County through low density zoning, but we believe that history will show you are relying on weak, inadequate protection; that Queen Anne's County will be unable to resist the pressures of growth which are operating on it; that development will occur and that you will have forsaken an important opportunity to move Queen Anne's County to the forefront of high quality, rational, controlled waterfront development. We believe you are looking too closely at "bad development" of the past and not enough at rational visions of good development which are available to you.

He cited predictions that the Baltimore–Washington area would add nearly 100,000 persons per year to its population for the next twenty years, and that this population growth was bound to continue affecting the Eastern Shore. Rouse's statement closed with this:

Low density zoning has been widely proven to be inadequate protection against the pressures of growth. Land zoned for large residential acreage consumes more valuable farm land and open space than it should. It is often later broken up into smaller parcels and many of its natural amenities are forever destroyed. At first one point and then another, zoning crumbles under pressure, and each break in the zoning wall leads to further breaks. Commercial uses bite at the edges of the highways and waterways and spread their ugly unplanned clutter. New kinds of large scale, well planned, good development are needed to provide new images of the possible and to erect new standards to undergird stronger planning and zoning requirements and to make real to the people of a community that there are decent alternatives to sprawl and clutter and the ravaging of the rivers. We care deeply about

Queen Anne's County and the Eastern Shore. We are enormously grateful for the openness and warmth of the hundreds of people with whom we have discussed our plans for Wye Island. We regret the circumstances which compel our withdrawal—both economic and governmental.

On May 29, 1974, Wye Island did not look a bit different than it had when the Rouse Company had entered into the option with the Hardys over a year before. No grading had been commenced, nor a dime spent for construction. For that matter, the Rouse Company had not paid the Hardys directly for the privilege of holding the option; the company's obligations under the option agreement had been to conduct exhaustive environmental and economic studies and to turn these results over to Frank Hardy if the option went unexercised. The costs to the Rouse Company of this paper project had been considerable. At the annual shareholders' meeting the following fall, Jim Rouse was asked how much these and the other expenses of the Wye Island project had cost the Rouse Company. "Between $800,000 and $900,000," he answered.

———————cʌɔ———————

It's not the big projects that do the damage," Scott Ditch was saying, "it's the piecemeal proliferation." Ditch was understandably steamed up about the collapse of the Wye Island project. As he paced about his carpeted office at Columbia, he whacked the back of his chair with a ruler for emphasis. "How did Rockville Pike happen? Well, it didn't. It was unplanned." Rockville Pike is a highway leading out of Washington through the Maryland suburbs; it is the preeminent example of haphazard, cluttered, strip development. "You go in and ask the planners, 'Where will the used car lots go?' 'Oh, we don't have anything like that,' they say. They never realistically provide for what *will* occur. Nobody gets excited about the little guy opening his store and selling tobacco. Rockville Pike could have been planned thirty years ago to take care of all those uses that have made it such an ugly mess. It could have been landscaped and well thought out. But if you'd come in with a plan for a mile deep along that Pike, they would have lynched you. It just gradually eroded away. No one fights the little erosion. They take on any major project, though. Anything large scale they fight.

"There's something people can't comprehend in a place like Queen Anne's County. They can't comprehend a sound developer. They can't comprehend an honest one. There was just disbelief—a complete lack of credibility. They would say, 'We know what you'll do. You'll get started and sooner or later you'll lower your prices and pretty soon we'll have all that crap.' We saw a long and, probably, a losing battle. Even if you get the zoning approval, you don't have clear sailing ahead. They can be slow in issuing building permits. There's even an attitude down there of hoping to keep Queen Anne's County out of the newspapers. They don't want it discovered. They just don't think growth will happen. 'Nah,' they say, 'not on Wye Island.'

"Another thing we ran up against was the fear that we'd take control of the county. The officials simply never responded. They never said, 'Gee, those numbers are too low.' Nothing. Not a word. Just 'Thank you and good-bye.' "

Scott Ditch quit pacing and sat down. The steam seemed to go out of his voice. He turned in his chair and looked out the window. "There's nothing now to prevent Hardy from selling off the island in bits and pieces," he said quietly. "I would like to come back to Wye Island, say, fifty years from now, and see what they did with it."

People in Queen Anne's County don't want another Columbia. Besides, no one would have been willing to buy land under Rouse's plan because of the restrictions on private docks.

I think Rouse had a good plan, with some minor changes. They kept coming into the county commissioners with different plans, but the commissioners would never tell Rouse what they wanted. I don't blame him for leaving here. I would have, too.

Good riddance is what I say.

The county shouldn't a'bucked him. The county here could use that taxable base. That's just going against progress.

Who needs the sonofabitch?

———————∿———————

Edgar Bryan had changed his mind. He stood on his back porch and shook his head. "They've driven Jim Rouse out, and I think that's a bad thing. I'd ruther seen it stay in agriculture, but that's just my personal opinion. I think one of the real mistakes Rouse made was to throw it open to the public. When you let every Tom, Dick, and Harry, who have no interest whatsoever, come in. . . . I was at that meeting of the Bennett Point group at the Grasonville school, and there wasn't one real native shoreman there. They all like it over here, but they don't want anybody else over here. Those city people don't care about their selfishness."

Edgar Bryan waved toward distant Bennett Point. "All those new people down on Bennett Point tell me, 'Mr. Bryan, please don't sell that farm. . . we love it. . . why, we saw two deer there just the other day!'" Edgar Bryan grimaced. "We've incorporated the farm 'cause we can't keep it intact, the marginal profit is so small. It's a question of economics."

I'm very glad that Jim Rouse did not succeed. I never thought he could, financially. I don't think the county's ready for that kind of development.

I think Rouse ought to have developed Wye Island and cleaned it up. I thought some upper-class people would come down and make it a real nice place for Queen Anne's County.

I'm glad they fell flat on their faces. There are just a few places left like this island. This land produces good corn, and that would all have stopped and the money [would have been] lost to the economy.

That would have been awful to see five thousand homes down there.

———————∿———————

Jim Rouse slumped into the chair until he was almost horizontal with the floor. It was early evening. He had just left his office, bringing to his modest Columbia town house two bulging and battered briefcases crammed full of articles and papers to read. The leather hinge on one briefcase was badly split.

Rouse twirled the ice in his drink with his finger. "I'm just convinced that Queen Anne's County will be worse off, not better off," he said. "The personally discouraging thing to me is that I know we have a record for good results. There was a pervasive neutrality. . . not a high degree of concern. The local newspaper editor wrote a column soliciting letters about our proposal—they never got *one*." His optimism momentarily flickered. "I think eventually we could have won the zoning, had it not been for the economic picture. With patience, the fire in the opposition could have been removed."

Rouse spoke about the growing opposition to growth of any kind that was becoming so pervasive in Queen Anne's and Talbot counties, particularly among the city people who were retiring to their country places, particularly with regard to Wye Island. "I think it is a blind selfishness," he said. "It is failing to deal realistically with the world as it is, attempting to push things somewhere else rather than seeing that they're done well. I have no reason to think it'll change over there, unless they defile Wye Island. This business of people being willing to accept a gradual erosion of good land use, instead of good development, is a strange thing, indeed. It happens slowly though, so gradually. It never happens decisively. There's no point when you say, 'Wow, it happened!' When the bad occurs, it's such a gradual accretion."

Rouse rubbed his freckled forehead. Behind him on the wall were hung some portraits one of his sons had painted of college friends—they wore somber expressions. "The events of Watergate," Rouse said, "confirm what the kids have been saying: they want to act by their consciences. You have to get rid of hate, and the way to do it is simply by being honest and truthful. I believe people are basically good, kind, and generous and that there are rational solutions in life. It is only a matter of getting hold of the solution, not that solutions to problems are impossible. But there are more creative answers than *stop*."

On the mantle over the fireplace there was a small, slim book entitled *To Believe in God*. Pieces of orange construction paper were stuck between some of the pages. One of the passages so marked read, "To believe in God is to have the great faith that somewhere, someone's not stupid."

———— ⌒⌒ ————

The day after Rouse's pullout from Wye Island, a cold drizzle settled over the Eastern Shore. Julius Grollman sat behind his rolltop desk, his feet on the cash register, and clipped his fingernails with a pair of ten-inch scissors. "I thank him for the good news," he said of Rouse's statement. The phone rang and Grollman picked up the receiver. "He gave up. He gave up," he said into it. "He said we wouldn't do what he wanted us to do." A man who was standing at the counter swung around and looked with surprise at Grollman. "Rouse has give up on Wye Island, huh?"

Julius Grollman settled back in his chair. It was quiet in the store, except for the snicking of Grollman's scissors and the infrequent murmur of a car passing through Stevensville. Julius yawned. "Rather than leave things wide open, you've got to set some limits," he said. "We'd like to see orderly, steady growth—not cities being built or anything like that. Keep growth at 3 percent a year. Industry-wise, we would like to have clean industry, and high pay, though I guess that's asking for too much."

Grollman placed the scissors on the desk and rocked further back in the chair. "They gotta do something to preserve agriculture. . . to preserve *some* places *some*wheres. What is it from Washington to Rockville? It's all *city*, isn't it?"

When Rouse announced his company's withdrawal from Wye Island, the Hardy brothers issued a press release reaffirming that, despite Rouse's departure, they would proceed to dispose of Wye Island. I asked Julius Grollman whether he considered five-acre lots on Wye Island preferable to Rouse's plan, and he quickly answered: "One thousand more families wouldn't be good either."

A man walked in the store and asked whether Grollman had any chicken feeders. As Julius Grollman got up from his chair, he said, "You don't just change the zoning for Wye Island. Before you know it, you get *ten* Wye Islands and *ten* times the population."

———— ⌒⌒ ————

Thirty miles away in Easton another man sat upright behind his steel office desk and fumed. "You can't compete with ignorance," Frank Hardy said, "especially if they won't learn."

Frank Hardy's business offices are on the second floor of the Loyola Federal Savings and Loan building in Easton. On the walls are a number of photographs of his prize beef cattle and a few of the yachts he has raced. The gray, wet day did little to improve his spirits. "It's a damn shame to see all that money spent on people who don't want to listen, who don't even want to go off Kent Island to look at *good* development. It's a tragic set of circumstances. On the one hand you have an incredibly intelligent man in Rouse. He's a real visionary. I think it's a great blow to the Chesapeake Bay, an immense setback, not for me, but for all sensitive shoreline development, because if you don't have some decent development, you're just going to have the Boise Cascade boys roaming around.

"There are over thirty-five million people within a three hours' drive of Queen Anne's County. And they come up with those asinine statements at the county level that they will limit the population to five hundred per year. The ignorance here is kind of overpowering. What are they going to do next? Say no more Catholics or Jews? Make it another Dachau? What kind of crap is that? This isn't Tierra del Fuego!

"The overreaction in this county has been very detrimental. Better than 97 percent of the high school graduates leave Queen Anne's County permanently because there's nothing for them to do. The only reason they come back is to attend their parents' funerals. We don't even have a movie theater. I guarantee you, if you had Bennett Point down here in Talbot, it would be sold out. We've had dozens of people come down and fall in love with Bennett Point. Then they nose around Queen Anne's County. They ask if there are any private schools. There aren't. 'Where do I shop?' You don't. 'Is there a country club?' No. Sooner or later, the people of Queen Anne's, the merchants, *et cetera*, will see they're not getting any of the action, and then everyone will want a trailer park. That's why I say that in fifteen years Queen Anne's County will be a very undesirable place to live. The wall will have crumbled. The old people will have died. And today they can't pass along these old estates to their kids—the estate taxes will kill them. Most of the people who own these expensive waterfront estates are over sixty, and when they die no one is going to buy them for farms. They'll start busting them up.

"In retrospect, I sometimes wonder why I didn't just lay out Wye Island in five-acre lots in the first place. But I didn't because I believe in good development. I'm not saying I'm carrying any cross of Lorraine, either. I don't think you can even sell Wye Island in this money market. If you had a different political machine in Queen Anne's that would accept Rouse, hell, they'd have jumped at it. I think the five-acre thing has proved to be something that the people don't want. Wye Island is big enough to sink any one person. You just take into account the front-end costs. I think the best thing is just to sit on it. If it can't be *properly* developed, it *shouldn't* be developed. I suspect that in ten years it will be worth $25 million. I do happen to give a damn about Queen Anne's County, although I may not sound it."

Frank Hardy's frown lifted. He stood up. "I'm not hurt, though. I've still got my island." Then he smiled, and chuckled, "A helluva lot of people know about Wye Island today who a year ago had never *heard* of it."

———————ⵢ———————

One of those people was Pat Noonan, the young, bright president of the Nature Conservancy. The Conservancy is undoubtedly the largest and most active nonprofit, land-acquiring conservation organization in the country. It buys, or has donated to it, ranches, swamps, islands, and other pieces of the natural landscape in order to prevent their development. Whatever land the Conservancy doesn't retain, it then sells at cost, or donates, to the National Park Service, or other federal, state, or local conservation organizations. Often the Conservancy is drawn into a pitched battle between conservationists and a developer over a particular parcel of land. Other acquisitions result from methodical, painstaking negotiations that stretch over years. Still others are strictly targets of opportunity. Wye Island was just such a target.

To develop a more systematic strategy for using its funds, the Conservancy engaged the Smithsonian Institution for an ecological survey of the Chesapeake Bay. The survey was to identify, for later protection, the biologically rich, "highest-quality samples" of "dense life forms," "abundant gene pools," or significant numbers and types of rare and endangered plants and animals. The Smithsonian team did not include Wye Island on the list; Wye Island did not have sufficiently dense life-forms, or enough gene pools, or rare or endangered species. The scientists did conclude that the island might have the rare and endangered Delmarva fox squirrel, so they placed a small triangle on their Wye Island map. But, on the whole, Wye Island did not pass scientific and ecological muster.

While the Rouse project was still alive, Pat Noonan was called by Shep Krech and others along the Talbot side of the Wye who pleaded for the Conservancy to step in and buy the island should Rouse fail. When Rouse pulled out, Noonan and his staff decided that, gene pools or not, Wye Island was a natural for the Conservancy. It had a good wildlife habitat, "that magnificent view," was surrounded by moneyed people—the Houghtons, the Wymans, the Krechs, and du Ponts, for openers—who wanted Wye Island to remain unchanged, and for all practical purposes, there was only a single owner to negotiate with. But Wye Island would not come cheap. Frank Hardy was in no mood to make a charitable gift of the island, whatever the particular tax advantages. The sale price to the Rouse Company was to have been $8.75 million, but now, stung by the county's rejection of Rouse's plan, Hardy was talking in terms of $10 million.

Noonan's first instinct was to shake the money tree of the U.S. Fish and Wildlife Service in the Department of the Interior. For years there had been vague talk about turning the island into a refuge, so Noonan approached Nathaniel Reed, whose responsibilities as the assistant secretary for fish, wildlife, and parks included overseeing the national wildlife refuges. Reed knew about Wye Island. He had hunted waterfowl on the Shore for a number of years. And his boss, Rogers Morton, lived across from the island. Reed had the Fish and Wildlife Service survey the island to determine its refuge potential. As expected, Reed was told that Wye Island was all that it had been rumored to be as a wintering spot for geese. But wildlife refuges can present political problems. For years gunners in the South had scanned their vacant skies with growing frustration, as more and more geese cut short their southerly migrations at the Chesapeake Bay. In addition, Wye Island would be one of the most expensive refuges in the United States, and the bay area already had a number of state refuges, two national wildlife refuges, and hundreds of square miles of open water, coves, and farmland to accommodate even more geese. Interior said no.

Pat Noonan talked with Arthur Houghton. Would Houghton be willing to buy Wye Island or organize a purchase group? Arthur Houghton demurred and suggested that Noonan contact Peter Thompson, a wealthy stockbroker in Easton.

Peter Thompson had moved from upstate New York to a farm near Centaur Castle in 1946, hoping to combine gentleman farming with stockbroking. He and his wife also bought and restored an historic home in Easton, from which they now manage a busy brokerage business. Like most of his friends, Peter Thompson did not want to see Wye Island developed, by Rouse, or Hardy, or anyone else. Thompson enjoys the leisure of the Eastern Shore. "You've shot ducks, picnicked on the banks, and you don't want it spoiled," he said. One day he and Clarence Miles attended the Preakness, and after the races they had difficulty getting out of the parking lot because of the glut of other automobiles. A car pulled into line ahead of them. Thompson's immediate reaction was, "Don't let that sonofabitch *in!*" That is the philosophy, he said, of all who move to the Shore. Peter Thompson wanted to hold the line around Wye Island.

Thompson was sympathetic to Pat Noonan's proposal: the Conservancy would buy Hardy's interest in Wye Island, using the collective millions of the gentry there to form the "take-out" group, which would then donate its development rights in Wye Island to the state of Maryland. But the portfolios of the wealthy were shrinking with the stock market. Even at a purchase price of $5 or $6 million, Peter Thompson was unable to stir up a financial commitment from those who claimed that Wye Island should remain in its natural state. None of them wanted a park, none of them wanted a Rouse village, and all knew that Hardy would not sit on Wye Island forever. But now that it had come time for them to pay for their wall, the gentry backed off. And so did the Nature Conservancy.

Peter Thomson believed that it was because Frank Hardy wanted to be a multimillionaire that the Shore was being ruined for people like himself. "We don't feel it's fair for them to make that much money to ruin the Shore for us," he said.

"Is Frank Hardy acting any differently than your clients who want to make the most out of their investments?"

"Some are that way," he said. "Some have more than enough money, but they insist on making more. Those are the ones that usually get hurt."

———— ✺ ————

In mid-June, a few weeks after Rouse's project collapsed, the following advertisement began running in Eastern Shore newspapers and those in Washington and Baltimore:

ANNOUNCEMENT
WYE ISLAND SALE
SMALL WATERFRONT FARMS

PLANS ARE NOW COMPLETE FOR THE
SALE OF
35 SMALL WATERFRONT FARMS
FROM
15–70 ACRES
EXCELLENT FINANCING AVAILABLE TO
QUALIFIED BUYERS.
SALES WILL BE PRIOR TO OR AT
AUCTION
20 JULY 1974 2 P.M.
ON WYE ISLAND

FOR DETAILS CONTACT 643-5021 OR 643-6400, OR YOUR BROKER.
PRIOR SALES AND AUCTION SALES MUST EXCEED
TOTAL SALES PRICE TO BE ANNOUNCED AT 2 P.M.
AUCTION TIME 20 JULY 1974.

Frank Hardy was doing what he had said he would do: he was "disposing" of Wye Island, or at least a third of it—Wye Hall Farm.

A month earlier Frank Hardy had said, "I'm not in any mad rush to develop Wye Island, but it's economically impossible to farm land that's worth $5,000 an acre." He seemed to be in a rush now. It was costing him $25,000 a year just to hold the Wye Hall Farm option. He had to exercise Ruth Dorlon's option by the end of July or lose it. But that meant facing a purchase price of $2.5 million. For that, Hardy needed much more money than he presently had. Cutting Wye Hall Farm into a number of parcels all larger than fifteen acres—above the threshold of Queen Anne's subdivision ordinance and, thereby, the moratorium, he planned to sell them off and put up for auction those that went unsold. The preauction sales contracts would be subject to cancellation were Hardy able to get more for the land at the auction. Conversely, the bids on the auctioned parcels could be rejected should they be collectively less than a bid for the farm in its entirety. And all sales would be subject to cancellation if Hardy dropped the option, after determining that the sales would not cover both the costs of the farm and a healthy profit. People who wanted a piece of Wye Island could be encouraged to pay more before the auction because of the fear they might lose out to a higher bid under the tent. Frank Hardy had considered most of the angles.

———— ✺ ————

On Kent Island, in the middle of a wheat field near the Nichols Building, there sat a small brick building. At one end was the Stevensville Post Office, in the middle a laundromat, and at the other end the real estate office where Frank Hardy was doing his best to sell off Wye Hall Farm. The atmosphere inside was not exactly frenetic, but there was an observable excitement. Phil Beard, one of Hardy's salesmen, escorted three potential buyers—a middle aged couple and an older man with a shock of white hair—into a conference room. Phil Beard, freckled and wearing a red-and-white-striped shirt, exuded cheer and positive thinking. Floating through his mind at this moment was an elastic bubble encasing his three buyers. Phil Beard believed that buyers were looking for their bubble, that is, the amount of land that would provide privacy. For some, five acres would be an ample bubble, and for others, fifty acres was not enough. When told that all the Wye Hall parcels could be resubdivided into lots of five acres, one woman had waved Beard away. "That's too dense," she had told him—her bubble was much bigger. On a table in the center of the conference room was some of the expensive detritus from Jim Rouse's capsized vision of Wye Island: the section of the Rouse model, now Hardy's, depicting Wye Hall Farm with its meticulously laid-out "estate residences," their wooded and pastured shorelines, interior fields, hedgerows, and woods. Frank Hardy's plan for the farm was decidedly less elaborate: strips of red yarn, laid out like bones of a halibut, delineated his parcels from the road to the shoreline. Between each wooley property line, Hardy had placed a metal-rimmed paper disc, marked with the lot number. The older man, Mr. Adams, placed his hands on his hips and leaned far over the model. "Hmmm," he said.

The previous day Frank Hardy had described the preauction sales as "an incredible success. We sold Wye Hall mansion in an hour. . . Chaires has received over five hundred solid phone calls since the ad was placed." Hardy said that most of them were younger, wealthy doctors, lawyers, and other professional people, looking to speculate on a slice of Wye Island and then to subdivide it into five-acre lots for a fast, and fat, killing.

Phil Beard switched off the light and turned on a slide projector. He punched up the first slide, an evocative stretch of wooded shoreline. Click, click. "This is a freshwater pond on parcel 25. The black ducks love it." Click, click. Flocks of Canada geese lifting into the sky. Click, click.

"Which fifteen-acre plots are for sale?" asked Mr. Adams.

Phil Beard switched on the light. Mr. Adams flipped up the Polaroid shades that were clipped to his glasses. In the manner of a Reno croupier, Beard began snapping over the lot discs. "Well, I know this one is sold," he said. Snap. "And so is this one." Snap.

"Is there a standard price per acre?"

"No."

"Which is the least-expensive fifteen acre waterfront parcel?"

"This one. It's $95,000." The most expensive fifteen-acre parcel, costing $185,000, lay on a point across the river from Shep Krech's house and duck blind.

"And this can be subdivided into five-acre lots?"

Phil Beard's face assumed a childlike innocence. "Your purpose is to partition the land?"

"Yes," answered Mr. Adams, "we would be having probably three to five people join us in buying a parcel."

"Would you like to see Wye Island?"

The three looked at each other, shrugged, smiled, and said they would like to see Wye Island. Beard scooped up a set of keys, herded his speculators into a car, and drove off. Phil Beard had his work cut out for him on this day; it was sweltering, and a white haze, Washington's first pollution alert of the year, lay sultry and heavy over the entire Shore. It was not the kind of day to show off Wye Island as an escape from the city, as a speculator's dream.

Hung on the wall of the sales office were a few of the Rouse–WMRT maps of Wye Island—its "framed views" (trees bracketing an open shoreline) and "visual buffers" (trees blocking the view)—and the urban regions across the bay. Frank Hardy walked over to the regional map. Red dots the size of a quarter were stuck to it. Each represented the equivalent of ten thousand people. The Eastern Shore had very few dots. The areas of Washington and Baltimore were blotched almost solidly red.

Hardy tapped the map with his finger. "That's my favorite one. That map just tells the story. You know, Maryland now is about the fourth fastest growing state in population. Where are they going to go? And they talk about 3 percent growth!"

Hardy pressed his finger against the crimson swarm engulfing Washington and its suburbs and looked at Bill Chaires. "We ought to take a few of these off here, and add a few more here," he said, laughing and swirling his hand around the Shore.

———— ∾ ————

That afternoon I drove to Wye Island. Along the lane that bisected Wye Hall Farm, wooden stakes had been driven in at intervals corresponding to the newly subdivided lot lines. Strips of red plastic were tacked to the tops of the stakes. Some 316 years had elapsed since Thomas Bradnox had cleared the trees to plant tobacco on Cedar Bradnox. With little change since, Wye Island had oscillated between plantations and farms, between owners as imaginative as Judge Bordley and those as strange as the Stewarts, and tenants as persevering, if unmoneyed, as the Whitbys. Now Wye Island was to be subdivided for speculating physicians, dentists, and lawyers from the cities across the bay, and by them into smaller lots for other speculators and, in time, people who would build houses on the water. The plastic strips hung motionless.

Perry Blades, the manager for Wye Hall Farm sat in a chair beneath a silver maple in his backyard, contemplating eviction. Blades is tall, ruggedly handsome, and ruggedly built. Both his taciturn manner and his features faintly suggest the late Gary Cooper. He chews tobacco with subtlety. He wore a white work cap on which was printed MILFORD SURE CROP FERTILIZERS. A native of the Shore, Blades had moved to this house—one of the three habitable ones left on the island—twenty years ago. But he lived here only so long as his employer owned the farm, for, like Sam Whitby before him, Blades was a tenant of the farm, not an owner. And now that Hardy was about to become the owner of Wye Hall Farm, Perry Blades and his family were being displaced, in part because empty Wye Hall mansion swallows a hundred gallons of expensive fuel oil on an icy winter day.

Perry Blades has farmed all his life. For years he resisted Ruth Dorlon's offers to take a vacation. When he finally relented, Perry Blades packed his car, and he and his wife drove through Texas, fulfilling his childhood ambition to see the big ranch spreads and their arched wooden entrances. Every time he came to a ranch with a big sign over the gate, he got out and took a picture of it. Perry Blades took pride in the care with which he tilled the fields of Wye Hall Farm; the plows were not turned into the fields until the soil had thoroughly dried. He knows that a field tilled wet will turn cloddy, bake into concrete, and take years to get back to normal texture.

Blades was philosophical about leaving Wye Island. He had purchased a smaller house in Wye Mills. "It's no big thing to be leaving," he said. "Oh, I would have liked to stay, but it doesn't much matter where I live as long as I have a roof over my head and a job."

Earlier that day a young man had driven into Blades's yard, fairly bubbling with enthusiasm over a lot that one of Hardy's salesmen had shown him by boat. He had lived abroad and traveled the world, the young man told Blades, but he had never seen anything like Wye Island. He could not wait to buy his fifteen acres. Perry Blades had not given the young man his opinion of those fifteen acres, because he had not been asked. What Perry Blades thought about that particular piece, which he had farmed for almost twenty years, was that there was not much to it except a lot of marsh and mosquitoes.

Blades wondered how the buyers who had thoughts of living on Wye Island were going to get their houses built. "What I'm wondering is how they're even gonna get their construction equipment over here, and their lumber," he said, referring to the confined Wye Narrows bridge and its restricted load limits. "You can't even get a bulldozer over of any size. I'm wondering about a concrete truck. Seven cubic yards would be overloaded."

About Frank Hardy's designs for Wye Hall Farm, Perry Blades was mildly contemptuous. He figured that sooner or later all but Arthur Bryan's piece would be subdivided and sold, and that in the end there would be as many houses on Wye Island as Rouse had proposed.

"And it won't be as well protected," he added.

As I left Wye Island, there were about fifteen people strung along the narrows bridge, fishing and crabbing. A portly, middle-aged man sat in an aluminum lawn chair against the rail, tending his lines. He wore a businessman's hat and a white, short-sleeved dress shirt, open at the neck. He had come directly from his office in Baltimore.

One of Frank Hardy's sales boats, bedecked with buyers, roared under the bridge and up the narrows. The man had heard there were lots for sale on Wye Island. "What are they asking?" he asked.

I told him. He laughed and wrinkled his lips as if he had bitten into an apple seed. "That's too much for me," he said, turning to a twine that was beginning to change its angle to the water.

———— ✿ ————

A brown Mercedes-Benz rolled into the restaurant parking lot at Kent Narrows and stopped. Frank Hardy and Bill Chaires got out. It was mid-July, a few days before the auction. I asked how the sales were proceeding, and Bill Chaires punched his fist into the air in a Wye-power salute. "Great!" he said, not too convincingly.

In fact, some of the bubbled euphoria that had earlier enveloped Hardy's sales campaign was beginning to pop and vanish. A few prospective buyers—even one or two who had actually signed sales contracts—had taken the trouble to contact Bobby Price, the attorney for the planning commission, about their rights to resubdivide down to five-acre lots. Price told them of a rather obscure county policy that was to discourage people living on private roads from resubdividing their land into five-acre plots. All but four of Hardy's Wye Hall parcels were on private roads, or on no roads at all. When Hardy announced the forthcoming auction, the Queen Anne's County officials informed local newspapers of the policy, and the *Bay Times* (Kent Island) ran a front-page story. With the news of the county's toughening stand on resubdivision, his customers were beginning to bob and weave, and Frank Hardy was not happy.

Frank Hardy felt that the county officials were singling him out. They had turned a deaf ear to the Rouse plan, and now they were frustrating his sales effort on Wye Island. He had been telling potential buyers that they would have the right to resubdivide their acreage into lots of five acres, as the present zoning of the island provided, if they complied with all county regulations.

"They are playing around with our property rights!" Frank Hardy boomed. "Hell, that's what our Constitution was founded on—*property rights!* They just try and interfere with Wye Island, and they're going to have one helluva lawsuit on their hands." He held up a list of the thirty-five lots on Wye Hall Farm; with a felt pen he had drawn a line through twelve of them, indicating signed sales contracts. "That island's zoned for five acres. How can they decide to change the zoning just on their whim?"

Earlier, Frank Hardy had said he was not maintaining that his was the best plan for Wye Island. The best plan, he said, was leaning against the wall in his office: the Rouse plan.

———————cΛɔ———————

Down the road, in his comfortable niche behind the counter, Julius Grollman reached over his shoulder and found a package of Juicy Fruit. He unwrapped a stick and rolled it into his mouth. About the forthcoming auction that weekend, Grollman gave a muffled chuckle. "Oh, he's just getting back at us," he said. Julius Grollman was not surprised that, compared with the Rouse plan, there was so little public opposition to Hardy's subdivision on Wye Island. Rouse would have brought in almost nine hundred families.

"Hardy isn't asking to change anything," Julius Grollman said. "Rouse would have had to change the whole zoning."

Hardy was just doing what a lot of developers were doing on the Eastern Shore: selling lots to people who had no intention of moving there, who only wanted to resell for a nice profit. Speculators don't become neighbors. They don't build houses. At least not right away.

———————cΛɔ———————

Land satisfies many human desires: beauty, shelter, privacy, territory, security. And because owning it involves much of one's savings and income, most people also view their land—their home, their farm, their summer cabin, their tract in the mountains—as an investment.

But no one, including Hardy's lot buyers and those who buy and sell land as a business, wants to be called a speculator. Speculator is an evil word. It conjures images of sharp-eyed people in sharkskin suits who fly by night.

"Do you consider yourself a speculator?" I asked the driver, as the Cadillac whizzed across Kent Island.

"No," he replied, civilly but firmly.

One might assume that he would be shadowy, evasive, and sparing of information about the world of land speculation. He is not. He is open, direct, almost sunny, and spilling over with facts about his business. Wide shoulders, long legs, a loping stride, crackling energy, staccato speech, and sports coats that are never dull. He refers to himself as a "nickel-and-dime operator," but Lester Carpenter Leonard, Jr., is one of the biggest land syndicators on the Eastern Shore.

"We're in this for the long haul. I am what you call the poor man's syndicator—two to three thousand dollars per investor."

When Leonard has a commitment from seven or eight investors—mostly doctors, dentists, lawyers, executives with money—he approaches a farmer and offers to buy his farm. Occasionally he will sweeten the pot with a little of his own cash, but that is not how Lester Leonard makes his living syndicating land. He gets a percentage when the sales contract is signed, continuing commissions for managing the syndicate, additional commissions when shares are sold or purchased, and a final commission when his group sells the farm. Although none of his syndicated Shore farms have been sold, Lester Leonard does a considerable business among the individual investors who trade in and out of the syndicates.

He keeps the names of his investors off the legal instruments of record, usually recording the deeds in the names of strawmen. "Why disclose who's in the property? You get a domestic furor, and you have a real mess trying to sell the land." Rarely does Leonard's name appear on the recorded deed, and only then as trustee for the straw-named syndicate; he does not want to start getting a reputation around the Eastern Shore of being a big operator. It is part of Lester Leonard's strategy of keeping a "low profile." He says, "You mention the word 'speculator' and he's a crook. You get too big and you build up a little resentment. You make a lot of money and people are jealous. You go to a farmer, and he says, 'Oh yeah, Lester Leonard, I know who he is.' It gets pretty difficult, so we use straws mostly."

Lester Leonard's low profile casts anything but a small shadow. The car he drives to the Shore each week from his office in Washington, DC, is a white Cadillac Eldorado with a bright red interior and a hood as long as a boat oar. "The only thing I spend money on is my car. I get a new one every year." In addition to his yearly Cadillac he has a couple of Ford Torinos and a Dodge Colt ("for the energy crisis"). Each year he trades them in for new models of the exact color and style in hopes that his neighbors will not notice that he is buying new cars every year. A conservative dresser he is not. "I love clothes," he told me. On the day he drove me around his Eastern Shore properties, Lester Leonard was outlined in red piping on a blue blazer that was divided by a wide tie spangled with reds, blues, and whites. "As a syndicator, you can't have domestic problems, you can't have scandal. You must be the epitome of respectability. You've got to watch getting intoxicated in public, too, so no hard liquor. You can't have fights. And yet you must be seen."

So Lester Leonard is seen around the Shore, as secretary of the Kent Island Yacht Club, and as a member of the local Moose, Lions, Veterans of Foreign Wars, Masons, and the American Legion. He had signed a contract to buy a six-figure house on Prospect Plantation so he would have "an appropriate home for Lester Leonard, when he comes down to the Shore and wants to entertain." He lives in "a very modest home" in College Park, in the Maryland suburbs of Washington, and owns an efficiency condominium in Annapolis and one in Ocean City. In this way, he said, he can unload one of the condominiums if times get tough, and no one will know it; if he owned a big expensive home and had to unload that, everyone would know that Lester Leonard is in hard times. And not only does Leonard not want his profile showing if he gets in financial trouble, he has no desire to relive the boom-and-bust years of his youth.

Lester Carpenter Leonard, Jr., was born forty-eight years ago in New Jersey and grew up in the luxury of servants, gardeners, boats, "nice cars," and an education at Choate. But his father, a successful trial lawyer, lost his wealth to illness, and Lester's comfortable surroundings vanished. Lester took any job he could—heavy construction, the Monmouth racetrack, anything. After the navy, college, and night law school, he opened an office on Lafayette Square across from the White House, and assembled his first small syndicate. "Buy a house, fix it up, and sell it," was his early pattern. Leonard kept his investments small—"nothing big," $25,000 houses, just a few buyers investing a little cash for appreciation. He branched out to Harpers Ferry and moved, for the first time, into raw "vacation" land.

By 1967 Lester Leonard was managing syndicates in everything from two-story walk-ups to rolling forest and suburban pastures—but he worried about the future. Land prices in the Washington area had risen so high that it was increasingly difficult to amass enough cash for the big turnovers, and Leonard judged that future increases would not bring enough return to warrant the risk. "I saw the handwriting on the wall: moratoriums, high prices, wouldn't be able to sell." He searched for a place where land values were low but had potential for soaring. On a state road map, using College Park as the center, he drew a circle the equivalent of an hour's drive from home. The arc cut through Queen Anne's County on the Eastern Shore, where inland farmland could then be purchased for about $500 an acre. He hooked up with a local real estate broker and began buying and syndicating property. Within eight years Lester Leonard had over a hundred syndicates in Queen Anne's County—over five thousand acres of farmland and about seventy-five buildings in the villages.

The Cadillac purred past Dave Nichols's memorials of the Kent Island subdivision boom. Leonard turned down a lane through wide cornfields to the edge of Eastern Bay. It was Shipping Creek Farm, once known as the J. Lemuel Bryan Farm, picked up in 1936 by the president of Carter Carburetor and now one of Leonard's farm syndicates. "This is the best and largest one I put together," he said. "The barns are in good shape, the roofs are new." It had taken about thirty investors to purchase these 223 acres for $425,000, a third of it cash. Each year the syndicate earns about $8,000 income from the farm, about an equal amount from the gunning lease and farmland rental, and around $500 for a small pier and boat slips. "There's the old tenant house," Leonard said, pointing to a small frame dwelling. "We get $167.50 a month for it."

The Cadillac swung out the lane and up the highway. Near Stevensville, Leonard turned into a subdivision—Bay City—and parked at the edge of the Chesapeake Bay, facing the bay bridge. Here in Dave Nichols's burst bubble, Leonard had bought one lot as an investment for himself and one for a syndicate. On the way out of Bay City, we passed some houses being constructed in the woods away from the water and a view of the bay.

"Why anyone would build a house in the woods here is beyond me," said Leonard.

Lester Leonard hates to admit that some of his investors have never bothered to examine their property before investing in it. They just ask him for a copy of the contract showing their share of the interest, taxes, other expenses, and income. "There is risk in everything," says Leonard. "I say to my investors, 'Is this your last three thousand? If it is, I don't want it.' I run my syndicates with a dictatorial hand. I have to. When they say, 'Joe Smith says you could have bought Shipping Creek for a lot less,' I tell 'em, 'You can get someone else.' You don't have meetings with them either, because then everyone comes in with a drink under his belt and you get disagreement. When a farmer is ready to sell, you don't tell him you've got to meet with your investors next Thursday. He wants that money *now*. I send letters out advising them when to send their checks and so on. Mostly phone calls. I do rule 'em with an iron hand. If they're the experts, fine, let them run it." Leonard admits to being equally unyielding with sellers and tenants. "If he thinks he sees a hole, he'll take you for a ride. You have to be tough. If you go to settlement and the seller decides he doesn't like the terms, sue him that day. *That day!*" Leonard gets his share of malicious damage—he has had tenants purposely let the water overflow in the upstairs tub and run down the walls and ceiling below. When that happens, he sues. That day.

———— ❧ ————

To a tour of the Eastern Shore Lester Leonard brings neither the birded eye of a naturalist nor the historian's fascination for events and dates. But a grove of trees, an obscure house, a farm all get a new identity as Leonard's oral calculator spills out his financial perspective. A cornfield: "Here's a farm we bought for $500 an acre. . . 216 acres. . . sold the old house for $35,000. . . this farm's now worth about $1,000 to $1,200 an acre." A stand of woods alongside the highway loomed ahead. He owned it. "Fifteen hundred an acre. . . everybody laughed at me," laughed Leonard. A farm outside of Centreville: "Do y'see all the road frontage this piece has? *Tremendous* road frontage. One of the reasons we purchased it." Leonard hoped to convince the town of Centreville to annex the farm and send out sewer and water lines. He would then subdivide or sell the farm as developable land. At the edge of Queenstown, Leonard pulled up to a brick-shaped building about the size of eight little bedrooms shoved together. Eight little spartan bedrooms was what it was—the Starlighter Motel. It is where Leonard stays when he is on the Shore. He owns it. He bought it for about $16,000, spent $9,000 to fix it up, and from it now nets about $3,600 a year. Twelve dollars a night for a single. Leonard pointed across the road to a nondescript building that, in his dreams, would have become a pizza and sub sandwich shop, a "twenty-four-hour deal," had the owner not stalled around for so long. Down Piney Neck to a farm near the roaring bulldozers at Prospect Plantation. One of Leonard's first Shore syndicates: "Waterfront. . . about $1,100 an acre. . . a real steal. This farm *conservatively* would sell for $300,000, but I think you could get $350,000. Y'see, we're smack-dab against Prospect Plantation." Only after buying the farm did Lester Leonard learn that his syndicate had purchased not from a farmer, but from another speculating syndicate. From College Park.

In Centreville, we continued the tour on foot. Leonard's syndicates make more in shorter periods of time on rental buildings than they do from farms. He pointed to a two-story brick building. "Would you believe $20,000? Twenty! How's *that?*" A triumphant smile. "Slate roof. Brick. It'll be here until the town burns down." I thought he was about to sell it to me. "Twenty thousand simoleons! We get about $300 a month rent." He ran his gaze from the sidewalk to the roof, then turned and winked: "Not bad." We walked down the street and stopped in front of a clapboard house. "Asbestos shingle. Ok, we get a $150 a month. Water and sewer. Cost? Twelve thousand, five hundred."

Leonard stepped into a small liquor store. Like a painter using his thumb for perspective, he squinted at a strip of lottery tickets, which he held at arm's length, and the list of winning numbers posted over the counter. No winners. He bought five dollars' worth of new tickets and walked out, striding across the street to one of his favorite "collecting" shops. When he is not checking his farms and buildings, Lester Leonard is collecting. He collects everything—gold, silver, coins, large boat models. (Once he saw an old gasoline pump, the type with the glass container on top, and he bought that.) To Leonard, collecting is part investment and part mania. The proprietor recognized Leonard and immediately produced a silver vase, which he said was very nice. Leonard examined it for about five seconds and said he would take it. Almost absentmindedly, after he began to fumble for his money, did Leonard ask the shopowner how much the vase cost. Thirty dollars. "Wrap it up."

Lester Leonard's horizons are ever expanding. Part of the explanation for this is purely a matter of making more money. Inland farmland in Queen Anne's County that sold for around $500 an acre when Leonard began syndicating farms there in 1967 was now selling for roughly $1,100 to $1,200. Waterfront farms that he had picked up eight years ago for $1,500 to $2,000 an acre were now costing roughly $5,000 an acre. So Lester Leonard would bypass the Hardy auction of Wye Hall Farm. Although the values of most of his farm syndicates had doubled, Leonard did not anticipate anything like another doubling in the next six or seven years. And it was now twice as expensive to put together farm syndicates on the Shore, more so, considering the costs to borrow money. Hardy had already got to Wye Island during the "opportunity period," and Leonard was not about to pay top dollar to syndicate land that might not double again in a half-dozen years. His horizons now stretched beyond Maryland's Eastern Shore to the northeastern wedge of the Delmarva Peninsula—to Delaware. There he expected to find cheaper farmland to buy, to hold, to resell.

The other part of the answer lies in Lester C. Leonard, Jr.'s vagabond restlessness. He is a man on the go, figuratively and literally. "I'm one of those people who have to have a lot of things going. I gotta be on the move, have to have a number of deals going to be happy." He handles a law practice in addition to his syndications (which include more than two-dozen buildings in Washington), but his cluttered, dim walk-up office in Washington, DC, suggested that Lester Leonard is a much happier man behind the wheel of his Eldorado than he is behind his desk. The entrance to Leonard's office was at the top of a dark stairway through a peepholed door quilted with black vinyl. It had all the gloomy aspect of a movie-set private eye's office. The ceiling leaked. His desk was a sea of checks, letters, newspapers, an empty Courvoisier VSOP carton; and spilled across a worn maroon carpet were subdivision maps, carpet swatches, dead potted plants, phone books, law books, and stacks of correspondence.

The phone rang. Lester began rummaging through the pockets of his green sports coat and yellow slacks. Out came a fistful of business cards. "Wait a minute." He stuck the cards back and pulled out a wad of receipts.

"You're going too fast." Into another pocket and out with a handful of crumpled currency. "I can't find it, but it's here somewhere. Call you back."

His mobile office on the Shore, unlike the walk-up in Washington, is immaculate red upholstery and a padded dash, and as Leonard drives around his farms and visits his village buildings, he simply dictates his observations into a cassette tape recorder. No papers, no phones, no bookwork.

"I don't like to stick to one thing for very long," Leonard said. And then out of nowhere he said he had been thinking of becoming an Episcopal minister. "Whether I'm humble enough to do it, I don't know, but the economy will make the difference." His face broke into a broad smile: "I don't want to be a *poor* minister." When he contemplated his new horizons in Delaware, the clerical vision dimmed. "I guess I'm too restless to be a minister."

Lester Leonard is also probably too restless to be a developer, but he has certain ideas about how the upper Shore should look in the future. Through his syndicates Leonard is one of the biggest landowners in Queen Anne's County, and his decisions on the disposition of that land will directly influence the upper Shore's fate. "To me, the Rouse proposal for Wye Island was beautiful; I was really in favor of it. I think the county will live to regret it." But of Kent Island, where he has at least four farm syndicates, Leonard expressed a somewhat different vision, clouded by an ambivalence of where to draw the line on development.

"I want to see development of Kent Island, high rises from the bay bridge up a lovely four-lane highway to Love Point. Perhaps low rise on the Chester River side. . . and single-family houses over the south half of the island. I'm talking about condos, not rentals. I want to see a mix, though, not just millionaires."

Lester Leonard paused, pondering the universal sentiment that attracts people to rural areas. "That view is fantastic," he said of the bay. He paused. "But if you let in too much, you destroy it for everyone."

Saturday, July 20, 1974. Less than two months since the collapse of the Rouse project. Auction day for Wye Hall Farm. The weather is on the side of the Hardy brothers. By noon the sun is warm, and sharp, cooling breezes are making the bay, and the hundreds of yachts and sailboats that have flooded across it, dance. The sky is a child's painting of primary blue. Since early morning the bay bridge and its approaches have been jammed with hundreds of cars and boat trailers heading for the ocean and the landings on the Eastern Shore. A station wagon is sitting out the wait behind the tollgates. Crammed into it are folding chairs, two bikes, a bushel basket, cots, blankets, two coolers, a woman, four children, and a furry black dog of immense size. A long-handled crab net is attached to the roof.

The narrows bridge to Wye Island is deserted, but cars and pickups are parked at each end. A young, rotund boy, wearing a gray athletic T-shirt, faded overalls, and a crownless yellow cap, stands on the bridge. He clutch-es an empty bushel basket. He is from Glen Burnie, a suburb of Baltimore. The state police have evicted him and the other crabbers from the bridge. His black face is downcast.

"There are signs that say no fishing or crabbing from the bridge, and today's the one day they have to enforce it," he says. "I only got about one-dozen crabs before that trooper run us off. I was hoping to get a bushel basket full." He throws the basket into his pickup truck and drives away.

At the gate leading into the Duck House, two of Frank Hardy's farmhands are sitting on chairs beneath a beach umbrella. "You're the first money," says the one who accepts my five dollars, the admission fee to the auction. "I'm gonna frame that."

Parked incongruously on the grass next to Grason Chance's abandoned farmhouse is a shiny helicopter. Its blades droop like the antennae of a gi-ant cricket. The field in front of the Duck House is filled by a huge marquee striped in green, white, and blue. Under its canvas are rows of folding chairs facing a wooden platform.

Frank Hardy paces around the grass parking lot, his tie snapping in the wind. Two cars drive in and park. "Joe Jackson will put on a pretty good show," says Hardy of his auctioneer. He looks around at the few cars that have arrived. "I think we'll do well today."

That morning Hardy had passed the crabbers on the narrows bridge. He didn't want any interference when his buyers began arriving, so he called in the state trooper to run them off. "But they just come back out of the bush-es," he says. "You can have some real problems with those chickenneckers. They can get pretty irate. Two years ago one of them got so mad he smashed in this guy's car window with one of those crab forks. Have you ever seen

195

those tongs they use for lifting out crabs?" The forks are over a foot long and solid metal. "He just beat in this guy's window."

A Dodge pickup arrives and backs up to a smaller marquee. Two men lift beer kegs wrapped in damp canvas out of the pickup and set them under the tent behind serving tables. A group of women and children, arranging plates and utensils for the lunch, bustle back and forth from the smaller tent to the Duck House kitchen, where other women are cooking crab cakes, a delicacy of the Eastern Shore. Hardy walks over to another pickup and takes a can of beer from a cooler.

The wind picks up paper plates and skitters them across the grass. It blows a sheet of smoke, pungent with the aroma of barbecuing chicken, under the serving tent. The barbecue operation is the domain of perspiring men of the Queenstown Volunteer Fire Department. They are experts at barbecuing chicken, which they do every Sunday afternoon at a local church. The bed of coals is contained in an iron trough, supported at one end by spoked iron wheels and at the other by a heavy tongue jacked up on blocks of wood. A red-and-white fire truck is parked close by. Today the firemen will cook 250 quarters of chicken.

Small trees and lower limbs of the large oaks that were growing behind the Duck House have been cut down and shoved off the embankment into Granary Creek. It is not an effort at shoreline control; Frank Hardy wants the bidders to enjoy an unobstructed view of the water during their lunch among the oaks. A few small boats head out of Granary Creek into the Wye East River. A large, stout man walks from table to table with a staple gun, trying, with limited success, to keep the red-checked tablecloths from being blown into the cove by the freshening wind. The man is Albert Greaves, Frank Hardy's farm manager. His father farmed Wye Island years ago when Albert was a teenager and ducks were thick on the Wye. He is sorry to see Wye Island being subdivided, by his boss or anyone else. "I hate to see it cut up into little chunks," he says, "but I guess that's the way it's all gonna go eventually. But I would prefer seeing it go in one chunk." Albert Greaves, stapling as he goes, weaves his way among couples who stand with drinks in their hands, examining the Rouse maps of Wye Island that are nailed to the trees. A bearded man, who looks somewhat out of place in a ballooning Hawaiian shirt, shorts, and sandals, has brought along his two boys. They are bored with the maps and begin to fidget. One pulls off his cap. The father jerks him around by the arm—"Now *look*, I don't want to have any of that! If you're not going to wear it. . ." Trays of chicken are now moving from the iron wagon to the serving tent. The chicken lunch costs three dollars, the crab cakes four dollars.

The tranquility of the day is shattered by the roar of the helicopters—there are now two. The whine becomes a grinding roar. One helicopter swings over the Duck House, its blades flapping loudly, the red light under its belly blinking off and on. Phil Beard is aloft with two buyers, who peer down on the moving panorama of the island. The pilot pushes the chopper forward, and it sweeps up the road and over the narrows bridge. Below, the crabbers are returning from the bushes. Up the narrows to Wye Landing. It is jammed with cars and trailers. All along the Wye East River outboards are scattered about like white chips, and around them bob hundreds of white and fluorescent-orange crab pot markers. The pilot banks the chopper and brings it back over the cornfields, low over Perry Blade's soon-to-be-vacated home, and lower still over the trees and the Duck House to the landing patch. The second helicopter is lifting off with another group of buyers.

William A. Hardy, Jr., Frank's brother, has flown up from Florida for the auction. He is hefty, wide of face, deeply tanned, and smoking a cigar. He stands under the striped marquee, comfortably dressed in a blue blazer, an open-necked shirt, tan trousers, brown loafers and no socks—much like the customers stepping from their Cadillacs and Mercedes, a uniform casualness. Bill Hardy removes his dark glasses. He says that he and his brother bought Wye Island as a long-term investment. He likes Florida. He talks about the alligators, which he respects, but prefers to see some distance away in the marsh, or on your shoes as leather. "They'll grab your bass plug so fast you can't see their mouths open and shut. They're really mean. They'll come right up out of the water. . . they can outrun a dog." Bill Hardy acts as relaxed as Frank. He leans back in a chair, links his hands behind his head, and talks about Florida. But his eyes constantly sweep the growing gallery of map watchers.

Bill Chaires, Hardy's sales manager, is floating under the marquee, looking right and left and trying to determine which of his hottest prospects have shown up. He is upbeat. "We've got some good ones," he says. Suddenly he spots a familiar face, a customer who has been undecided about buying. Bill Chaires bursts into effusive goodwill, raising his arm in a wave.

"There's the good doctor!" he announces with sufficient volume to carry well beyond the good doctor. But just as suddenly, Chaires realizes that the good doctor's name has momentarily vanished from his memory. "I. . . damn," he mutters, "I can't remember that name." Chaires halts his salute in mid-air, brushes back his hair and, turning away from the nameless customer as if he is investigating something in the woods, Bill Chaires crabs sideways for the Duck House.

One-thirty. Cars continue to arrive, but not fast enough for Frank and Bill Hardy. They are glancing more often now at the parking area, assessing the rate at which it is filling. A bar has been laid out at the rear of the auction tent, and it is now doing a brisk business. Joe Jackson, the auctioneer, has arrived. Tall and lanky, he steps onto the platform and announces into the public address system that all who intend to bid must register at the Duck House and get their bidding numbers. The auction had been scheduled to begin at 2:00 p.m. Jackson announces that the bidding will be delayed for half an hour, because the traffic is backed up on the bay bridge.

Seated alone in the front row is a serious man who is making notes. He glances up from time to time at the subdivision map tacked to the platform railing. He is Tom Wyman, owner of Wye Heights, the splendorous estate that overlooks the east end of Wye Hall Farm from across the river. Native bison and exotic deer roam within his fences. Like the du Ponts, the Krechs, and others along the Wye East River who share a view of Wye Island, Tom Wyman had opposed the Rouse project. "The Rouse plan looked okay at the outset, and I suppose visually it would have been better because of no docks, and so on, but it would have never stayed that way," he says. But Wyman had not wanted even the initial plan because "*that* would have meant almost three thousand people." Wyman has lived on the Eastern Shore for about thirty years, rather he commutes from New York to the Shore for weekends. "They talk about a thirty-minute back up on the bay bridge—Christ, back then you could hardly *get* here from New York." From his commanding bluffs he has watched the boats on the Wye increase in number, particularly those of the chickenneckers who trailer down to nearby Wye Landing. "Jesus," he says, "this morning when I came out, all those crabbers had put out about 250 to 300 of those little white plastic floats fastened to their crab pots. Not 20 or 30, but 250 to 300! They don't contribute a goddamn thing to this county."

Wyman continues to mark down the parcel numbers and prices of the lots that face his estate. He intends to bid on them, to buy his view of Wye Island. If the price is right.

The seats under the striped awning are filling up. Frank Hardy has changed to red-and-white-checked slacks. He walks through the aisles and offers sharpened pencils from a cigar box. The whine from the helicopter sounds like a dentist's drill, as the chopper lifts off and whirls over the auction grounds. Joe Jackson reminds the crowd that they must register at the Duck House for a bidding number if they intend to buy. "The selling will start as soon as we can get that helicopter down." Mr. and Mrs. Eugene du Pont III have settled into the sixth row, a subdivision map of the farm spread out on the grass at their feet, Eugene leans down and points with his pencil to a lot, then consults the price list that he holds in his other hand.

At two-thirty David Bryan, the Hardys' attorney, steps onto the platform. He is Edgar Bryan's son, Arthur's nephew. His secretary seats herself behind a calculating machine placed on a metal desk. Frank Hardy reaches over the railing and punches a few keys. Bryan announces that because of the level of preauction sales, the Hardy brothers, Frabil Partnership, have decided to exercise the option to buy Wye Hall Farm. The unsold lots will be auctioned off first, then the larger assemblages, and finally the entire farm. The Hardys, says Bryan, reserve the right to reject any and all bids. Hunting will not be allowed on the farm for the coming season, and percolation tests will have to await next spring.

Joe Jackson moves to the microphone. People shift in their chairs. "Make your checks payable to the Wye Hall Farm Escrow Account," says Jackson. He snaps a card in his hand. "Are there any questions? Are you sure there are no questions? Because we want to get into the selling!"

"Will the current county moratorium prevent any of us from subdividing the parcels we buy?"

The moratorium applies to any proposed subdivision of five or more lots, David Bryan says, and the minimum lot size for Wye Island is five acres. The crowd mumbles, computes—any parcel under twenty-five acres slips under the moratorium. There are twenty-two such parcels to be auctioned.

Joe Jackson leaps back to the microphone, the arm holding the lot card stretched straight out in front of him and the other held high over his head. He looks like a charioteer. "OK, here we go now! Tract number *two* is the first one we are selling today!" The field that was once part of the planta-tion of William Paca, signer of the Declaration of Independence, that Sam Whitby had known as the Mansfield place, where Perry Blades has sown the last sweet corn, is now going on the auction block as three subdivided parcels. Parcel one was sold prior to the auction. The bidders lean forward. Jackson's sharp voice rivets the air.

"Wha'd'ya say? How much d'ya say? Four thousand an acre, do I hear four? Four, do I hear four, four, four, four?" A bid at $3,000 an acre. "Will you give thirty-one, will you give thirty-one, one, one, one, will you give thirty-one, one, one, one?" No one will give $3,100 an acre. Three Baltimore physicians hold fast to their $3,000 bid. Jackson keeps stabbing the air for $3,100, but there is no movement under the marquee. A piece of the Mans-field place goes to Baltimore. A brief pause for the bidding number and price, and Joe Jackson takes off again. The auction is underway.

The lots are consecutively numbered from parcel number one, at the bridge, clockwise around the entire farm. Joe Jackson works his way along Wye Narrows.

"All right, this is tract number four. How much? Start it, *please!*" There is an uncomfortable silence. Frank Hardy casually looks around, slowly fold-ing his arms. He sits about halfway back, blending in with the bidders.

"Will you give $3,000 an acre to start it?" Silence. "Will you give three, three, three, three, three, will you give *three?* Jackson's words fly. "How much will you give, what'll y'give, will you give $3,000 a bid, will you give three, three, three?" His tone firms, ever so slightly, suggesting he has hit rock bottom. No further. "Will you give *three?*" Jackson waits. His two spot-ters work their way through the audience looking for bidders, for hungry faces. "Twenty-five-hundred-dollar bid. Will anybody want twenty-five, five, five, five, five, *five?*" Jackson repeats the lower number, the cadence of his voice rocking, "Two thousand-dollar bid, two, two, two, *two.*" Jackson looks almost defiant now. "Will you give $2,000?" There must be some mistake. "Now this is tract number four," says Jackson, "look at all that waterfront!" Joe Jackson pleads, *"Anyone?"* For fifteen seconds the only sound heard is the clicking of ice in plastic cups and the wind fluttering the tents. Fifteen seconds of silence during the bidding in an auction seems like an eternity. Finally a head nods. *"Thank* you!" Jackson is off again, "Twenty-one, one, one, twenty-one, twenty-one, one, one, one, twenty-one?" Silence. Stillness.

Jackson's practiced senses tell him that the bigger quarry are not to be found for tract number four, but he squeezes out the last bit of drama, the last bit of anxiety. "Two thousand dollars *once.*" Pause. "Two thousand dollars *twice?*" Jackson's voice carries the hint of a fortune slipping away. He pauses for ten seconds. The tent is hushed. The spotters freeze. Jackson relaxes his tense frame. "Your number, please?" he asks of the man with the one and only bid.

The audience also begins to relax, but Jackson's chariot is off and pounding around the curve. "All right, tract number five! Anybody start it at $4,000, four, four, four, four? Come on, let's *go!* Four thousand, four, four, four, four thousand, *four!* Will someone give thirty-five, five, five, *five?*" A sense of unease is beginning to creep around the tent. "*Look,*" Joe Jackson shouts, jabbing his hands at the crowd, "you're buying *Wye Island!*" Two men in polo shirts, summer trousers, and sockless loafers stand up, pull their chairs from under the shaded marquee, and sit down again, drinks in their hands. They both stretch out and close their eyes against the warming sun. The Hardy auction is in trouble.

"Twenty-five, five, five, five, will you give twenty-five, five, five? *Anybody* want to bid on tract number five?" asks Jackson. A man waves his bid card and says, "I don't want to bid twenty-five, but I'll bid lower." Joe Jackson asks how much. "Fifteen," says the man. "Well, sir," says Joe Jackson, "I'm afraid, I'm afraid you are wasting your time." "Fifteen," says the man again, but Jackson ignores the bid. Jackson is also wasting his time. Tract number five is passed up.

Frank Hardy does not unfold his arms. He sits very still, looking straight ahead at the platform, not smiling, not showing any expression. He gives no outward sign of discomfort, but Frank Hardy is not comfortable. He is asking $184,000 for the forty-six acres of parcel two—the bid is $138,000. Parcel four raises only a single bid, for $126,000, just over half what Hardy had hoped for. And nobody is willing to bid on number five, except the man whose bid Joe Jackson has rejected because it is only a little over a third of Hardy's asking price.

Joe Jackson tries his skills at humor. "You know, ladies and gentlemen, they say there are three things you should consider when you buy land. The *first* is location. The *second* is location. And the *third* is location!" There is a tinkle of laughter that the wind blows away. "I think Wye Island is a real purty place," says Jackson.

The chariot is off again. "Make me an offer!! Two thousand to start it. *Anybody!* Will you give two, two, two, will you give *two?*" No one will give two or anything else. Tract number six is passed by.

Around the Wye Narrows comes Joe Jackson. Parcel twelve. "All right, start it out! Now remember, you have 1,950 feet on the water. Will you give $10,000 to start it?" Silence. "Eight! Will you give $8,000, eight, eight, eight?" Silence. "Seven, Seven, seven, seven, will you give *seven* to start it?" Silence. Jackson grips his hips, exasperated. "Will you give *five?* Will *anybody* give five?"

"I'll give five," says a voice.

"Thank you, sir," says Jackson, "fifty-one?"

Parcel twelve is one-half of the eastern tip of the farm, its woods visible from Tom Wyman's estate. Tom Wyman nods.

The spotters pivot and home in on the other bidder and Tom Wyman. The man nods for $5,200 an acre. "I've got *two!* Good!" shouts Jackson. "I've got *two*, will you give three, three, three, three?"

Jackson stares at Wyman. The spotters stare at Wyman. Wyman can imagine a boat dock sticking into the river from parcel twelve. Wyman nods a bid, and Jackson immediately swings to face the other bidder.

"Four, four, four. . . fifty-four?" The man does not nod. A long silence. Jackson's shoulders slump, and he accepts Tom Wyman's bid. A few people straggle off toward the portable toilets. The members of the Queenstown Volunteer Fire Department douse the coals in the barbecue wagon, sending a wet-smelling smoke across the bidding grounds.

Jackson is galloping down the Wye East River now. "Come on in and *buy*! Now nine, nine, nine, nine, w'd'ya go nine, nine, nine, *nine?*"

One of the spotters, a teller in the Centreville bank, sees a man widen his eyes and turn to talk to his wife. The teller closes in. "Don't talk to *her*, talk to *me!*" says Joe Jackson and the audience laughs, but a cautious wife snuffs out the bid.

"Twenty-eight hundred dollars, *once!*" Pause. "Twenty-eight hundred dollars *twice!*" Jackson hesitates in posed anticipation. A spotter catches a slight movement of a head and snaps his hand up. "I've got twenty-*nine!*"

yells Jackson, hurtling off and up. "Thirty-one, one, one, thirty-one now, *one* now, *one* now, *one*, thirty-one, thirty-one, thirty-one, now *one*."

Both spotters now hover over a group of five men who have drawn their chairs together in a circle. The men's heads nearly touch, and the sound of their muffled debate is just barely audible. Jackson bores in: "Do you want another *board* meeting?" he asks, evoking laughter. Frank Hardy sits directly behind the huddle, arms folded. The spotters practically join the meeting. Finally, one of the men nods.

"Sold!" bellows Joe Jackson. "Sir, could I have your bidder's number?" The men fall back in their chairs—smiling somewhat nervously—and reach down for their drinks.

Phil Beard drifts along the periphery of the auction tent. He looks discouraged. Bill Chaires no longer emits exuberance. The tension and excitement that characterized the opening of the auction have dissipated. Anxiety shows only in the faces of a few people who walk nervously around the bar at the rear of the marquee. Most of them are buyers who signed sales contracts before the auction, hoping to get a piece of Wye Island at bargain prices. A physician from the Washington suburbs makes no attempt to disguise his chagrin. When he signed his contract, he was told by Hardy's salesman that he was lucky to get in then because the auction prices later would be even higher. But now the bidding is telling the physician that he had contracted to pay twice as much as he might have at the auction.

"All right, give me a *bid!* Five thousand-dollar bid! Will you give five, five, five. Twenty five hundred dollars twice! *Third* and last call. Remember, you're buying *Wye Island*. It'll be the last chance you have at public sale. Where *are* you?"

"Three thousand to bid. Will you give three, three, three? Two thousand to start it. Anybody go *two*? Your bid's mine; make me an *offer!*"

"Tract twenty-two. It faces southeast. Eight thousand dollars an acre and *go!*" Seven. Five. Four. "I've got *four!* Don't lose this one."

"Tract twenty-three. Will you give me seven? Six thousand? Will you give me five? Five, five, five. Think of those beautiful sunsets you'll look at! Will you start it at five? *Four* thousand! That's as low as we go." Pause. "I've got *three!!* Will you give thirty-one, one, one, one." The bidding moves up, then freezes at $4,600 an acre. A spotter circles three men who huddle together, their harsh whispers revealing a subdued panic. One says he will bid

$4,700, "Sold!" Frank Hardy had valued this point across the river from Shep Krech's farm at almost $1,100 an acre.

When the bidding on the last parcel ends, Joe Jackson begins reoffering them in groups of larger acreage. "Will you give two, two, two, will you give $2,000 an acre?" Jackson pleads, but no one will. Nine combinations are offered. No bids. It is late in the afternoon, the wind has fallen off, and the heat of July oozes everywhere. Empty bottles are piled behind the bar. The torpor infects even Jackson, whose words come with less velocity now.

Among the stretched-out bidders and onlookers only one man is afire: a real estate speculator from the Western Shore. He is built like a wrestler gone soft, but his eyes are those of a man possessed. He races from row to row, slapping a folded map in his hand and pleading with groups of no-longer-huddled bidders to join him to buy a group of lots. He is racing for the lifeboats, and no one will pull him aboard. He is frantic. He pounds the map in front of Frank Hardy's face. "Come *on*, Frank, we'll put it together for thirty-*four!*" Frank Hardy looks with disgust into the man's florid face and without comment walks toward the Duck House.

"Come *on*, Bill!" the man shouts into Bill Chaires's face, and Bill Chaires turns away. The man looks pleadingly at Phil Beard. Beard shrugs his shoulders and shakes his head. The man is now gasping, almost apoplectic. He rushes over to a clutch of physicians who had been bidding earlier, "Come on!" he shouts, slapping the map against his thigh. "Come *ooonnnnn!*" Joe Jackson announces that bidding on the last collection of parcels has ended. His mouth agape in stunned defeat, the man opens his arms and turns his palms and face skyward, as if to tell God that he could have got it wholesale.

Jackson announces that Wye Hall Farm in its entirety will now be offered for sale. He states the minimum bid, the collective amount of auction bids and preauction contracts: $3,087,078.85. "Price per acre is $3,468.63," he says. Jackson is only going through the motions. "Thirty-four sixty-eight once? Thirty-four sixty-eight twice?" With little conviction Jackson asks, "Any bids?" The bidding is over.

The Hardy brothers and David Bryan retire to the Duck House for a ten-minute conference. They return to the bidding platform, and David Bryan steps to the microphone and says: "Ladies and gentlemen, on behalf of Mr. Bill Hardy and Mr. Frank Hardy, I would just like to say that it is a beautiful day this afternoon on beautiful Wye Island, and the bar is still open. I am sorry to announce that all bids are rejected. Thank you all very much for coming and have a nice trip home. Your checks will be returned at the river entrance to the gun club." The preauction buyers sigh in relief—they still have their lots. Bill Chaires's secretary walks up to three of them and says, "Well, do you feel better now?" They nod, let the air out of their lungs and say, "Whew!"

It is about 5:30. Most of the cars are gone, and only a handful of people are standing around. David Bryan pours himself a drink at the littered bar, as Frank Hardy walks up.

"Well, Frank, the tent proved to be big enough," Bryan says.

"It's not a tent," Hardy says, smiling thinly, "it's a marquee."

AFTERWORD

It is the fall of 1976, more than two years after the aborted auction of Wye Hall Farm. One of the preauction lot buyers is building a home, Frank Hardy leveled another abandoned farmhouse, and vandals burned down the old Bryan house where Sam Whitby grew up. Other than those abrasions, Wye Island looks little different than it did when the auction marquee was up.

Its ownership, however, is about to change. The state of Maryland has contracted to buy out the Hardy interest in Wye Island for $5.3 million. Frank Hardy had hoped to hang onto Wye Hall Farm, resubdivide and sell it off into five-acre lots (the moratorium has been lifted), leaving to the state the remaining 1,700 acres or so of the island. He had these negotiations just about wrapped up when the Queen Anne's County commissioners raised a fuss about partial acquisition. In order to spread its funds further, the state lowered its negotiating price per acre and finally signed a contract with the Hardys to buy all the island (all, that is, except Wye Hall mansion, the seven lots Hardy has sold, and the Bryan place which Arthur Bryan will not sell, short of condemnation proceedings the state is reluctant to begin). Getting all the money may still be a problem, even though the U.S. Department of the Interior is throwing in $2.1 million from the Land and Water Conservation Fund (the second largest grant to Maryland in the history of the fund). But if the state ultimately acquires Wye Island, this much is certain: it will be little used by the public. People on the Shore are dead set against a park. Maryland intends only to lease the island for farming—to the Hardys for the first two years—and allow some goose hunting. But no landings, no campgrounds. No chickenneckers.

Over three centuries ago Lord Baltimore encouraged settlement of Maryland by giving away chunks of the colony to tobacco planters. Now the state of Maryland is buying back Wye Island to keep the new settlers out.

———— ✎ ————

Historically in this country, communities marked their progress by their new houses, stores, and roads. Today, however, more and more residents look upon new development as a threat to their community's stability and their own personal happiness. So they fight it, and often successfully so. I have tried in this book to examine why they do.

To protect the environment? There is much of that, particularly the esthetic one. Fear of higher taxes? That also. But there is more to this discontent than concern, real or imagined, over potential ecological abuse and higher taxes. It is the resistance to change brought on by new people moving—or threatening to move—in. The desire of people to exclude others is what fueled the opposition to the Rouse project and the state park for Wye Island, and it is what much of the fight over growth is about throughout this country.

Many dispute this observation. People are opposing development, they say, in a rational effort to protect the "carrying capacity" of the land. But carrying capacity is usually a highly subjective standard. As the Rouse project demonstrated, if homes are clustered and central water and sewerage facilities are built, even sensitive environments can accommodate a great number of people—a greater number, however, than most are willing to accept. The westward movement was one of Americans wanting to get there first; today they want to be the last ones in.

The line between insiders and outsiders, between the excluders and the excluded, is a fluid one at best. And the rules by which people should be permitted to close off their communities to others are yet to be decided. Pitted in opposition are two fundamental prerogatives that Americans rightly cherish: to keep one's neighborhood familiar and unchanging, and to improve one's life by moving on. The resolution of these conflicts, therefore, is less a matter of determining natural and physical limits of the environment, than it is a balancing of human aspirations and values. And about that we have much to learn.

ABOUT THE AUTHOR

As the first secretary of the newly established Council on Environmental Quality and then Deputy Under Secretary of the Interior in the Nixon Administration, Boyd Gibbons was instrumental in helping shape important White House land-use initiatives, including new environmental regulations on strip mining, controls on stream channelization, and changes in tax laws to protect wetlands, restore historic buildings, and encourage charitable land transfers for conservation. He was a leader in the administration's 1970 decision to halt plans for the development of an international jetport in the Big Cypress Swamp, a key part of the Everglades National Park watershed.

From 1997 to 2006, Gibbons was president of the Johnson Foundation, a nonprofit organization that sponsors Wingspread conferences to encourage candid dialogue on a range of civic and community issues, including school reform and sustainable development. He was Director of the California Department of Fish and Game from 1992 to 1995. Formerly a senior research associate at Resources for the Future and a legislative assistant in the United States Senate, he was a member of the senior editorial staff of *National Geographic* for 15 years. He has written on a variety of subjects for the magazine and has published two books of non-fiction: *Wye Island* and *The Retriever Game*. Gibbons holds bachelor and law degrees from the University of Arizona.